Reviews of Environmental Contamination and Toxicology

VOLUME 180

Springer
New York
Berlin
Heidelberg
Hong Kong
London
Milan
Paris
Tokyo

Reviews of Environmental Contamination and Toxicology

Continuation of Residue Reviews

Editor
George W. Ware

Editorial Board
Lilia A. Albert, Xalapa, Veracruz, Mexico
D.G. Crosby, Davis, California, USA · Pim de Voogt, Amsterdam, The Netherlands
O. Hutzinger, Bayreuth, Germany · James B. Knaak, Getzville, NY, USA
Foster L. Mayer, Gulf Breeze, Florida, USA · D.P. Morgan, Cedar Rapids, Iowa, USA
Douglas L. Park, Washington DC, USA · Ronald S. Tjeerdema, Davis, California, USA
Raymond S.H. Yang, Fort Collins, Colorado, USA

Founding Editor
Francis A. Gunther

VOLUME 180

Springer

Coordinating Board of Editors

DR. GEORGE W. WARE, *Editor*
Reviews of Environmental Contamination and Toxicology

5794 E. Camino del Celador
Tucson, Arizona 85750, USA
(520) 299-3735 (phone and FAX)

DR. HERBERT N. NIGG, *Editor*
Bulletin of Environmental Contamination and Toxicology

University of Florida
700 Experimental Station Road
Lake Alfred, Florida 33850, USA
(941) 956-1151; FAX (941) 956-4631

DR. DANIEL R. DOERGE, *Editor*
Archives of Environmental Contamination and Toxicology

6022 Southwind Drive
N. Little Rock, Arkansas, 72118, USA
(870) 791-3555; FAX (870) 791-2499

Springer-Verlag
New York: 175 Fifth Avenue, New York, NY 10010, USA
Heidelberg: Postfach 10 52 80, 69042 Heidelberg, Germany

Library of Congress Catalog Card Number 62-18595.
Printed in the United States of America.

ISSN 0179-5953

Printed on acid-free paper.

© 2004 Springer-Verlag New York, Inc.
All rights reserved. This work may not be translated or copied in whole or in part without the written permission of the publisher (Springer-Verlag New York, Inc., 175 Fifth Avenue, New York, NY 10010, USA), except for brief excerpts in connection with reviews or scholarly analysis. Use in connection with any form of information storage and retrieval, electronic adaptation, computer software, or by similar or dissimilar methodology now known or hereafter developed is forbidden. The use in this publication of trade names, trademarks, service marks, and similar terms, even if they are not identified as such, is not to be taken as an expression of opinion as to whether or not they are subject to proprietary rights.

Printed in the United States of America.

ISBN 0-387-40402-3   SPIN 10936788

www.springer-ny.com

Springer-Verlag   New York Berlin Heidelberg
A member of BertelsmannSpringer Science+Business Media GmbH

Foreword

International concern in scientific, industrial, and governmental communities over traces of xenobiotics in foods and in both abiotic and biotic environments has justified the present triumvirate of specialized publications in this field: comprehensive reviews, rapidly published research papers and progress reports, and archival documentations. These three international publications are integrated and scheduled to provide the coherency essential for nonduplicative and current progress in a field as dynamic and complex as environmental contamination and toxicology. This series is reserved exclusively for the diversified literature on "toxic" chemicals in our food, our feeds, our homes, recreational and working surroundings, our domestic animals, our wildlife and ourselves. Tremendous efforts worldwide have been mobilized to evaluate the nature, presence, magnitude, fate, and toxicology of the chemicals loosed upon the earth. Among the sequelae of this broad new emphasis is an undeniable need for an articulated set of authoritative publications, where one can find the latest important world literature produced by these emerging areas of science together with documentation of pertinent ancillary legislation.

Research directors and legislative or administrative advisers do not have the time to scan the escalating number of technical publications that may contain articles important to current responsibility. Rather, these individuals need the background provided by detailed reviews and the assurance that the latest information is made available to them, all with minimal literature searching. Similarly, the scientist assigned or attracted to a new problem is required to glean all literature pertinent to the task, to publish new developments or important new experimental details quickly, to inform others of findings that might alter their own efforts, and eventually to publish all his/her supporting data and conclusions for archival purposes.

In the fields of environmental contamination and toxicology, the sum of these concerns and responsibilities is decisively addressed by the uniform, encompassing, and timely publication format of the Springer-Verlag (Heidelberg and New York) triumvirate:

Reviews of Environmental Contamination and Toxicology [Vol. 1 through 97 (1962–1986) as Residue Reviews] for detailed review articles concerned with any aspects of chemical contaminants, including pesticides, in the total environment with toxicological considerations and consequences.

Bulletin of Environmental Contamination and Toxicology (Vol. 1 in 1966) for rapid publication of short reports of significant advances and discoveries in the fields of air, soil, water, and food contamination and pollution as well as methodology and other disciplines concerned with the introduction, presence, and effects of toxicants in the total environment.

Archives of Environmental Contamination and Toxicology (Vol.1 in 1973) for important complete articles emphasizing and describing original experimental or theoretical research work pertaining to the scientific aspects of chemical contaminants in the environment.

Manuscripts for *Reviews* and the *Archives* are in identical formats and are peer reviewed by scientists in the field for adequacy and value; manuscripts for the *Bulletin* are also reviewed, but are published by photo-offset from camera-ready copy to provide the latest results with minimum delay. The individual editors of these three publications comprise the joint Coordinating Board of Editors with referral within the Board of manuscripts submitted to one publication but deemed by major emphasis or length more suitable for one of the others.

Coordinating Board of Editors

Preface

Thanks to our news media, today's lay person may be familiar with such environmental topics as ozone depletion, global warming, greenhouse effect, nuclear and toxic waste disposal, massive marine oil spills, acid rain resulting from atmospheric SO_2 and NO_x, contamination of the marine commons, deforestation, radioactive leaks from nuclear power generators, free chlorine and CFC (chlorofluorocarbon) effects on the ozone layer, mad cow disease, pesticide residues in foods, green chemistry or green technology, volatile organic compounds (VOCs), hormone- or endocrine-disrupting chemicals, declining sperm counts, and immune system suppression by pesticides, just to cite a few. Some of the more current, and perhaps less familiar, additions include *xenobiotic transport, solute transport, Tiers 1 and 2, USEPA to cabinet status, and zero-discharge*. These are only the most prevalent topics of national interest. In more localized settings, residents are faced with leaking underground fuel tanks, movement of nitrates and industrial solvents into groundwater, air pollution and "stay-indoors" alerts in our major cities, radon seepage into homes, poor indoor air quality, chemical spills from overturned railroad tank cars, suspected health effects from living near high-voltage transmission lines, and food contamination by "flesh-eating" bacteria and other fungal or bacterial toxins.

It should then come as no surprise that the '90s generation is the first of mankind to have become afflicted with *chemophobia*, the pervasive and acute fear of chemicals.

There is abundant evidence, however, that virtually all organic chemicals are degraded or dissipated in our not-so-fragile environment, despite efforts by environmental ethicists and the media to persuade us otherwise. However, for most scientists involved in environmental contaminant reduction, there is indeed room for improvement in all spheres.

Environmentalism is the newest global political force, resulting in the emergence of multi-national consortia to control pollution and the evolution of the environmental ethic. Will the new politics of the 21st century be a consortium of technologists and environmentalists or a progressive confrontation? These matters are of genuine concern to governmental agencies and legislative bodies around the world, for many serious chemical incidents have resulted from accidents and improper use.

For those who make the decisions about how our planet is managed, there is an ongoing need for continual surveillance and intelligent controls to avoid endangering the environment, the public health, and wildlife. Ensuring safety-in-use of the many chemicals involved in our highly industrialized culture is a dynamic challenge, for the old, established materials are continually being displaced by newly developed molecules more acceptable to federal and state regulatory agencies, public health officials, and environmentalists.

Adequate safety-in-use evaluations of all chemicals persistent in our air, foodstuffs, and drinking water are not simple matters, and they incorporate the judgments of many individuals highly trained in a variety of complex biological, chemical, food technological, medical, pharmacological, and toxicological disciplines.

Reviews of Environmental Contamination and Toxicology continues to serve as an integrating factor both in focusing attention on those matters requiring further study and in collating for variously trained readers current knowledge in specific important areas involved with chemical contaminants in the total environment. Previous volumes of *Reviews* illustrate these objectives.

Because manuscripts are published in the order in which they are received in final form, it may seem that some important aspects of analytical chemistry, bioaccumulation, biochemistry, human and animal medicine, legislation, pharmacology, physiology, regulation, and toxicology have been neglected at times. However, these apparent omissions are recognized, and pertinent manuscripts are in preparation. The field is so very large and the interests in it are so varied that the Editor and the Editorial Board earnestly solicit authors and suggestions of underrepresented topics to make this international book series yet more useful and worthwhile.

Reviews of Environmental Contamination and Toxicology attempts to provide concise, critical reviews of timely advances, philosophy, and significant areas of accomplished or needed endeavor in the total field of xenobiotics in any segment of the environment, as well as toxicological implications. These reviews can be either general or specific, but properly they may lie in the domains of analytical chemistry and its methodology, biochemistry, human and animal medicine, legislation, pharmacology, physiology, regulation, and toxicology. Certain affairs in food technology concerned specifically with pesticide and other food-additive problems are also appropriate subjects.

Justification for the preparation of any review for this book series is that it deals with some aspect of the many real problems arising from the presence of any foreign chemical in our surroundings. Thus, manuscripts may encompass case studies from any country. Added plant or animal pest-control chemicals or their metabolites that may persist into food and animal feeds are within this scope. Food additives (substances deliberately added to foods for flavor, odor, appearance, and preservation, as well as those inadvertently added during manufacture, packing, distribution, and storage) are also considered suitable review material. Additionally, chemical contamination in any manner of air, water, soil, or plant or animal life is within these objectives and their purview.

Normally, manuscripts are contributed by invitation, but suggested topics are welcome. Preliminary communication with the Editor is recommended before volunteered review manuscripts are submitted.

Tucson, Arizona   G.W.W.

Table of Contents

Foreword v
Preface vii

Veterinary Medicines in the Environment 1
    A.B.A. BOXALL, L.A. FOGG, P.A. BLACKWELL, P. KAY,
    E.J. PEMBERTON, AND A. CROXFORD

Health Effects of *Acanthamoeba* spp. and its Potential for Waterborne
Transmission 93
    NENA NWACHUKU AND CHARLES P. GERBA

Arsenic Hazards to Humans, Plants, and Animals from Gold Mining 133
    RONALD EISLER

Cumulative and Comprehensive Subject Matter Index:
Volumes 171–180 167

Veterinary Medicines in the Environment

A.B.A. Boxall, L.A. Fogg, P.A. Blackwell, P. Kay,
E.J. Pemberton, and A. Croxford

Contents

I. Introduction	2
II. Environmental Assessment of Veterinary Medicines	2
A. Responsible Authorities	2
B. Environmental Risk Assessment in the European Union (EU)	3
C. Environmental Risk Assessment in the United States	3
D. Proposals of the Veterinary International Co-operation on Harmonisation (VICH)	5
E. Environmental Risk Assessment Models	6
III. Veterinary Medicine Use	9
A. Ectoparasiticides and Endectocides	9
B. Antibiotics	9
C. Endoparasiticides	11
D. Antifungals	12
E. Aquaculture	12
F. Hormones	12
G. Growth Promoters	12
H. Others	12
IV. Pathways to the Environment	13
A. Emissions During Manufacturing and Formulation	13
B. Aquaculture	13
C. Agriculture (Livestock Production)	14
D. Companion/Domestic Animals	16
E. Disposal of Unwanted Drugs	16
V. Occurrence in the Environment	17
A. Aquaculture	17
B. Agriculture	28
VI. Metabolism and Environmental Fate	31
A. Metabolism of Veterinary Medicines	32
B. Fate in Manure and Slurry	32
C. Fate in Soil	33
D. Fate in Surface Waters	34

Communicated by George W. Ware.

A.B.A. Boxall (✉) · L.A. Fogg · P.A. Blackwell · P. Kay
Cranfield Centre for EcoChemistry, Shardlow Hall, Shardlow, Derby, DE72 2GN, UK.

E.J. Pemberton · A. Croxford
Environment Agency, National Centre for Ecotoxicology and Hazardous Substances, Evenlode House, Howbery Park, Wallingford, Oxon, UK OX10 8BD.

E. Fate in Sediment	36
VII. Environmental Hazard	36
A. Aquatic Toxicity	43
B. Terrestrial Effects	71
C. Estrogenic Activity	72
VIII. Recommendations for Further Work	73
Summary	74
Appendix A.	76
Acknowledgments	81
References	82

I. Introduction

Veterinary medicines are widely used to treat disease and protect the health of animals. Dietary enhancing feed additives (growth promoters) are also incorporated into the feed of animals reared for food to improve their growth rates. Release of veterinary medicines to the environment occurs both directly, for example, the use of medicines in fish farms, and indirectly, via the application of animal manure containing excreted products to land. A number of groups of veterinary medicines, primarily sheep dip chemicals (Environment Agency 1998, 2000, 2001; SEPA 2000), fish farm medicines (Davies et al. 1998; Jacobsen and Berglind 1988), and anthelmintics (McCracken 1993; McKellar 1997; Ridsdill-Smith 1988; Strong 1993; Wall and Strong 1987) have been well studied. However, as there are scant data available in the public domain on the environmental fate, behavior, and effects of other generic groups of veterinary medicines, their potential environmental impacts are less well understood (Jørgensen and Halling-Sørensen 2000). This review was therefore performed to gain a greater understanding of the potential risks to the environment arising from the use of veterinary medicinal products. The review considers publicly available data on exposure routes, environmental fate, behavior, and effects of all generic groups of veterinary medicines. It is hoped that the outputs from this review will be used to target monitoring programs effectively, to ensure that appropriate risk management is in place, and steer future research initiatives. All veterinary medicines, drugs, and pesticides referred to in this review are detailed in Appendix A, including common name, CAS number, and complete chemical name.

II. Environmental Assessment of Veterinary Medicines
A. Responsible Authorities

A pharmaceutical company is required to demonstrate the quality, safety, and efficacy of a new pharmaceutical product before it can be marketed. In the United States (U.S.), the regulatory authority is the U.S. Food and Drug Administration (FDA). The equivalent authority in the European Union (EU) is the European Medicines Evaluation Authority (EMEA) for Europe-wide authorization or the Member State's regulatory authority if individual country authorization is sought. The approaches used by the EU and the U.S. along with international initiatives to harmonize approaches to assess the environmental risk of veterinary medicines are treated separately next.

B. Environmental Risk Assessment in the European Union (EU)

Under EU Directive 81/852/EEC as amended by Directive 92/18/EEC (EU 1992) it is necessary, when applying for marketing authorization for a veterinary product, to assess the potential harmful effects that use of the product may cause to the environment and identify any precautionary measures which may be necessary to reduce such risks. The assessments are performed in two phases [Committee for Veterinary Medicinal Products (CVMP) 1997].

In the first phase (Phase 1), the potential for exposure of the environment to the product, its ingredients, and relevant metabolites is assessed. If the Phase 1 assessment indicates that concentrations in dung excreted to pasture, manure that is subsequently applied to land, soil, and water, are likely to exceed 10 µg kg^{-1}, 100 µg kg^{-1}, 10 µg kg^{-1}, or 0.1 µg l^{-1}, respectively, then the second phase (Phase 2) is required. Phase 2 is performed in two parts, Tier A and Tier B.

In Tier A, the possible fate and effects of the drug and/or its major metabolites are assessed in more detail than in Phase 1. Depending on the characteristics of the drug, tests may be required on aquatic species (fish, daphnids, and algae), earthworms, and plants, and there may be a need to determine the degradation half-life of the active substance in the environmental compartment(s) of interest. Typical Tier A studies are shown in Fig. 1. If the product exhibits insecticidal activity, then additional studies on dung fauna (one species each of dung fly and dung beetle) and grassland invertebrates need to be assessed. If after Tier A there is an indication that the compound poses an environmental hazard and that any proposed risk management strategies are inadequate, then Tier B is required. Tier B involves the study of effects on the fauna and flora within the environmental compartments that are likely to be affected.

C. Environmental Risk Assessment in the United States

The U.S. Food and Drug Administration (FDA) is required under the National Environmental Policy Act (NEPA) to consider the environmental impacts of its actions, including investigational use of drugs and approvals of drugs. Assessments are performed by the FDA Center for Veterinary Medicines (CVM) using data on exposure and effects that are provided by the sponsor of a new animal drug. The approach used involves a customized data acquisition plan for each new animal drug and is performed in four phases:

1. Problem formulation. Problem formulation identifies the major factors to be considered in the assessment; this will identify certain general areas of concern and information required to satisfy regulatory agencies.
2. Obtain information/data. Two types of data are needed, namely, data on exposure of the environment to the drug or its biodegradation products and data on potential hazards. Exposure data identify whether the drug and/or its deg-

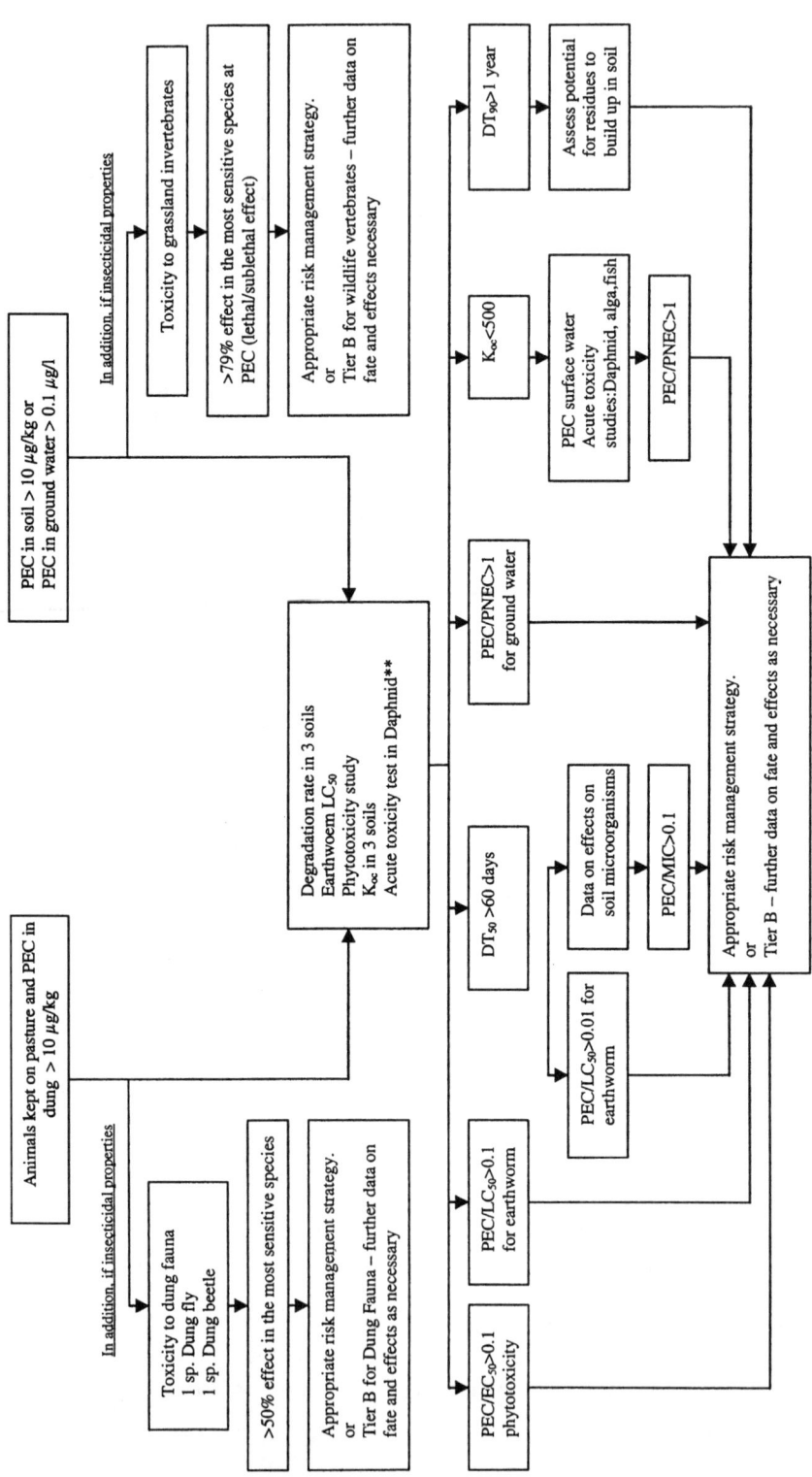

Fig. 1. Phase II—Tier A decision tree for medicines other than fish medicines (CVMP 1997).

radates are broken down by physical, chemical, or biological means, whether it evaporates to the air, remains dissolved into sediments, or will bind to sediment or soils. Information on animal husbandry, waste disposal practices, and the types of environment likely to be impacted may also be useful. Information on hazard is usually obtained from a series of standardized toxicity tests. The design of the test depends on the problems identified and the exposure data. For example, if a drug is water soluble and breaks down readily, acute toxicity testing on fish, daphnids, and algae is appropriate. If a drug is used regularly and persists, chronic testing may also be required.

3. Analysis of data. In the third step, both the exposure and ecological effects are characterized. The exposure characterization determines which environmental compartment(s) are likely to be exposed, and likely concentrations in these compartments are determined. The assessment may be quite generalized or focus on particular sites where sensitive or endangered species may be exposed. The ecological effects are characterized using the available toxicity test data in conjunction with appropriate uncertainty factors.

4. Risk characterization. The profiles of exposure and effects are the input to the risk estimation. Supporting information in the form of a weight of evidence argument may also be incorporated (e.g., data obtained from tests not meeting FDA quality standards). Estimated risks are then considered in terms of the types and magnitude of expected effects, the spatial and temporal extent of the effects, and the potential for the ecosystem to recover.

The ecological risk assessment is then incorporated into a public document known as an environmental assessment (EA). A number of these assessments can be downloaded from the FDA website (www.fda.gov/cvm/efoi/ea).

D. Proposals of the Veterinary International Co-operation on Harmonisation (VICH)

In 1996, the VICH Steering Committee authorized the formation of a working group to develop harmonized guidance for conducting environmental impact assessments for veterinary medicines in the EU, Japan, and the U.S. A guidance document on the Phase 1 assessment of veterinary medicinal products has been finalized (VICH 2000), and a number of guidance documents for Phase 2 are currently in draft form.

The Phase 1 guidance document uses a decision tree that allows applicants to conclude whether a product qualifies for a Phase 1 report (Fig. 2). Products that do not require a Phase 2 assessment include natural substances; products used only in nonfood animals; products used to treat a small number of animals in a flock or herd; products that are extensively metabolized (i.e., individual metabolites are excreted in amounts less than 5% of the applied dose); products used in aquaculture or on livestock that do not enter the aquatic environment because the treatment waste is disposed of by incineration or similar means; aquaculture products where the environmental introduction concentration is less than 1 µg l^{-1}; and products used on livestock that have a predicted soil concen-

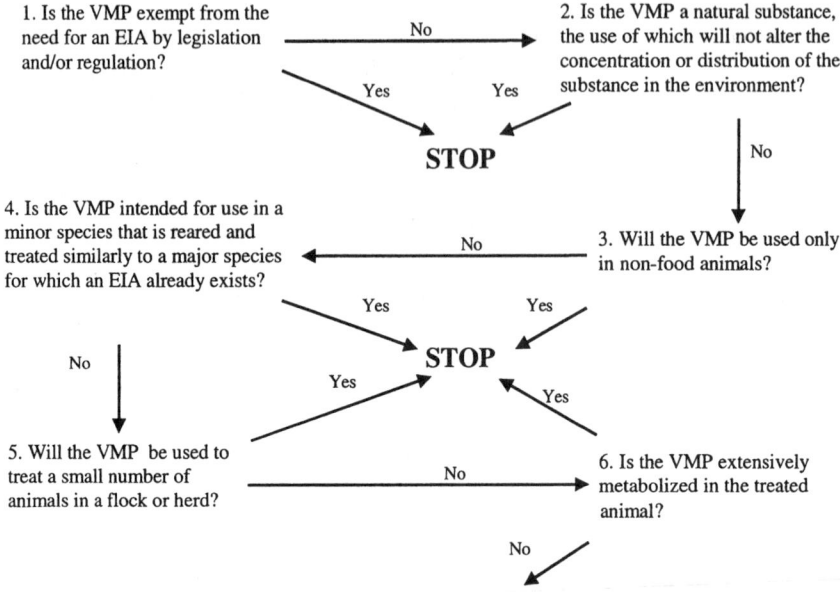

Fig. 2. Phase I decision tree (VICH 2000).
Continued on next page

tration less than 100 μg kg^{-1}. Draft Phase 2 guidance documents are available for products used in aquaculture, and for the treatment of intensively reared animals and the treatment of animals on pasture.

E. Environmental Risk Assessment Models

To support the environmental risk assessment process, a number of approaches have been developed for predicting concentrations of veterinary medicines in soil, groundwater, and surface waters (e.g., Montforts 1999; Spaepen et al. 1997; WRc-NSF 2000). An overview of each model is presented in Table 1 and discussed next. It should be noted that, because of a lack of monitoring data, none of the models described here has been validated.

Uniform Approach for Predicting Environmental Concentrations of Veterinary Medicines (Spaepen et al. 1997) To harmonize environmental assessments of veterinary products, the European Federation for Animal Health (FEDESA) developed a uniform scheme for calculating predicted environmental concentrations (Spaepen et al. 1997). The scheme provides a sequence of standard equations and a database containing information on three major agricultural animals: cattle, pigs, and poultry. The database also contains information on the agricultural practices and relevant regulations for various regions within the EU. Inputs to the model are the dose and treatment regimen. If information is available on

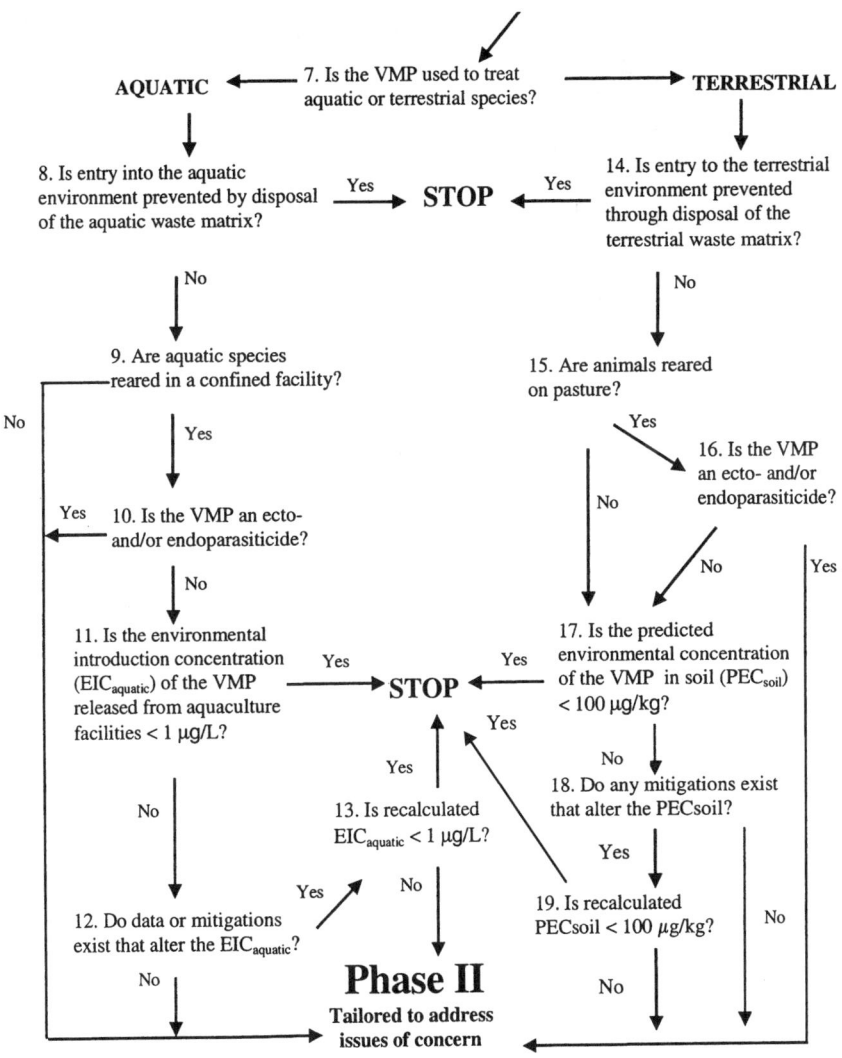

Fig. 2. (*Continued*).

metabolism or degradation, this can be incorporated into the calculation. The output from the model is a predicted soil concentration.

ETox (Montforts, 1999) The ETox model developed by Montforts (1999) predict concentrations of veterinary medicines using scenarios that are specific to agricultural practices in the Netherlands. The model is more complex than the uniform approach and can be used for medicines that are given internally (e.g., oral and injection treatments) or applied externally (e.g., udder disinfection treatments). A range of input pathways is considered (i.e., direct excretion of dung and urine onto a field; spreading of manure and slurry and direct spillage onto

Table 1. Exposure assessment models developed for use on veterinary medicines.

	Animals	Location	Husbandry	Administration route	User inputs	Outputs	Validation status
Uniform approach	Cows Pigs Poultry	Europe Member states	Intensively reared livestock	Internal treatment	Dose Treatment regimen Metabolism Degradation data	Concentration in slurry and soil	Not validated
Etox	Cows Poultry Pigs	The Netherlands	Intensively reared livestock Grazing animals	Internal and external treatments	Dose Treatment regimen Metabolism Degradation rate K_{ow}	Concentration in slurry, manure, soil, groundwater, surface waters, and biota	Not validated
VETPEC	Cows Pigs Poultry Sheep	Three UK catchments	Uses stocking densities at the county level	Internal treatment	Dose Treatment regimen Vapor pressure K_{oc} Molecular weight Solubility	Concentrations in soil, groundwater, and surface waters	Not validated

a field). The following groups of organisms are considered: cows (milk cows, suckling cows, beef cows), pigs (fattening pigs, sows), and poultry (hens, broilers and turkeys). The outputs from the model include concentrations of the veterinary drug in soil, groundwater, surface waters, and biota.

VETPEC (WRc-NSF) VETPEC is a combination of four existing models: a Mackay fugacity model for partitioning in soils; the PESTAQ and PESTCAT models for transport to groundwater and river water, respectively; and the uniform approach of Spaepen et al. (1997), which was just described. The model considers a range of animals, including cows, pigs, broiler chickens, laying hens, turkeys, and lambs and predicts concentrations in soil, groundwater, and surface waters in three UK catchments (the Cotswolds, Otter Valley, and Herefordshire). The model allows the user to produce outputs of likely concentration distributions that reflect the variability.

III. Veterinary Medicine Use

Only a limited amount of information is available on the usage of veterinary medicines (VMD 2001; Liddel 2000; Pellicaan et al. 2000). Very limited data are available for the U.S. and, with the exception of the UK, data are only available on the sales of antibacterials in individual EU member states. Data for the UK cover a wider range of products, including parasiticides, antimicrobials, and hormones. A summary of the available data is provided below. Although much of the information is UK focused, it is likely that this provides an indication of the types and relative amounts of different products used across the world.

A. Ectoparasiticides and Endectocides

Ectoparasiticides are used to control external parasites in livestock and endectocides are used to treat both internal and external parasites. Both ectoparasiticides and endectocides are used to treat parasites in a wide range of animals. If uncontrolled, ectoparasites (mites, blowfly, lice, ticks, headfly, and keds) can severely affect the welfare of sheep. In the UK, a number of product types are available and a range of active substances is approved for use (Table 2). The available data on the usage of ectoparasiticides on sheep (Liddel 2001; Pepper and Carter 2000) indicate that the organophosphate compound diazinon is the most widely used active substance, followed by the pyrethroids. Currently, only a limited amount of the macrocyclic lactones are used. Usage data of other ectoparasiticide active substances (i.e., phosmet, emamectin benzoate, piperonyl butoxide), used in agriculture, aquaculture, and treatment of companion animals, respectively, were not available.

B. Antibiotics

Antibiotics are used in the treatment and prevention of bacterial diseases (Gustafson and Bowen 1997). Their use follows principles similar to those used in human medicines, but there are some differences. The most significant difference is that livestock and poultry are raised in large numbers and it is therefore

Table 2. Major veterinary medicines in use in the UK.

Group	Chemical class	Major active ingredients
Ectoparasiticides	Organophosphates	Diazinon
	Synthetic pyrethroids	Flumethrin
		Cypermethrin
	Amidines	Amitraz
Antibiotics	Tetracyclines	Oxytetracycline
		Chlortetracycine
		Tetracycline
	Sulphonamides	Sulphadiazine
		Sulphadimidine
		Formosulphathiazole
	β-Lactams	Amoxicillin
		Procaine penicillin
		Procaine benzylpenicillin
	Aminoglycosides	Dihydrostreptomycin
		Neomycin
		Apramycin
	Macrolides	Tylosin
	Fluoroquinolones	Enrofloxacin
	2,4-Diaminopyrimidines	Trimethoprim
	Pleuromutilins	Tiamulin
	Lincosamides	Lincomycin
		Clyndamycin
Endectocides	Macrolide endectins	Ivermectin
		Doramectin
		Eprimomectin
	Pyrimidines	Pyrantel
		Morantel
	Benzamidazoles	Triclabendazole
		Fenbendazole
	Others	Levamisole
		Nitroxynil
Hormones		Altrenogest
		Progesterone
		Medroxyprogesterone
		Delmadinone
		Methyltestosterone
		Estradiol benzoate
		Ethenyl estradiol
Antifungals	Biguanide/gluconate	Chlorhexidine
	Azole	Miconazole
	Other	Griseofulvin
Anesthetics		Isoflurane
		Halothane
		Procaine
		Lido/lignocaine

Table 2. (*Continued*).

Group	Chemical class	Major active ingredients
Euthanasia products		Pentobarbitone
Analgesics		Metamyzole
Tranquilizers		Phenobarbitone
NSAIDs		Phenylbutazone
		Caprofen
Enteric bloat preparations		Dimethicone
		Poloxalene

necessary to treat the entire flock or herd at risk. In the UK and Netherlands, the available data (FIDIN 2000; Pelicaan et al. 2000; VMD 2001) indicate that in 1998 the tetracyclines were the most widely used antibacterial medicines, followed by potentiated sulfonamides, β-lactams, macrolides, aminoglycosides, fluoroquinolones, and others. Sales data for Norway for 1996 (antibiotics for therapeutic use in domestic animals and farmed fish only) showed the sulfonamides and aminoglycosides to be the major compounds used (Grave et al. 1999). The major active substances used as antibiotics are shown in Table 2.

C. Endoparasiticides

Endoparasiticides, which are used to control internal parasites, include anthelmintics (wormers) for the control of gastrointestinal worms, lungworms, and flukes as well as antiprotozoals and coccidiostats, which are included in feeding stuffs mainly for therapeutic or prophylactic purposes (Bowen 1995).

A wide range of active ingredients is used to treat gastrointestinal worms, liver fluke, and lung worms in poultry, cattle, sheep, and horses. The available data (IMS 2000) indicate that ivermectin, a macrolide endectin, is sold in the highest amounts (although this may include the sheep treatments already described), followed by the pyrimidines, azoles, and nitroxynil. Data available for North America indicate that approximately half the U.S. cattle population is treated with an anthelmintic and that approximately 50% of these treatments contain ivermectin (Forbes 1993).

Coccidiostats and antiprotozoals are often incorporated into feed stuffs for medicinal purposes, including prophylactic use for the prevention of diseases such as coccidiosis and swine dysentry and therapeutic use for the treatment of diseases. Apart from one individual substance (dimetridazole), usage data are largely unavailable (IMS 2000). The following compounds are considered to be potential major usage compounds within the therapeutic group: amprolium, clopidol, lasalocid acid, maduramicin, narasin, nicarbazin, and robenidine hydrochloride. In addition, major usage protozoal compounds include toltrazuril, decoquinate, and diclazuril.

D. Antifungals

Antifungal agents are used topically and orally to treat fungal and yeast infections. The most common uses include treatment of ringworm and yeast infections. The publicly available data indicate that the major active substances used are chlorhexidine, miconazole, and griseofulvin.

E. Aquaculture

A range of substances is used in aquaculture to treat mainly sea lice infestations and furunculosis. The medicines may be applied by injection, in feed, or via cage treatments. Substances used include oxytetracycline, oxolinic acid, amoxycillin, cotrimazine, florfenicol, sarafloxacin, emamectin benzoate, cypermethrin, teflubenzuron, azamethiphos, and hydrogen peroxide.

F. Hormones

Now banned in the EU as growth promoters, hormones have other uses, including induction of ovulatory estrus, suppression of estrus, systemic progesterone therapy, and treatment of hypersexuality. The major active substances used were altrenogest and progesterone.

G. Growth Promoters

Growth promoters, also called digestive enhancers, are antibiotic compounds added to animal feed stuffs to improve the efficiency of food digestion. From 1993 to 1998, sales of antimicrobial growth promoters remained largely static. However, in 1999, sales fell by 69%. This decrease is considered to be due to the ban by the EU in mid-1999 of those growth promoters that may confer cross-resistance to antimicrobials in human medicine (VMD 2001). Usage data on individual antimicrobial compounds used as growth promoters are limited. Sales data for Denmark for 1995 show tylosin and olaquindox as the two compounds sold in the greatest quantities for use as growth promoters (Jørgensen and Halling-Sørensen 2000). Compounds identified as potentially major usage growth promoters include monensin, flavophospolipol, and salinomycin sodium.

H. Others

Several other therapeutic groups that are used as veterinary medicines in significant quantities include anesthetics, euthanasia products, analgesics, tranquilizers, nonsteroidal antiinflammatory drugs (NSAIDS), and enteric preparations. The major usage compounds for each therapeutic group are listed in Table 2. In addition to these, the following "other" therapeutic groups have also been identified as potentially important: antiseptics, steroids, diuretics, cardiovascular and respiratory treatments, locomotor treatments, and immunological products. However, insufficient information was available to identify individual compounds and usage within each of these groups.

IV. Pathways to the Environment

Veterinary medicines enter the environment by a number of different pathways. Emissions may occur at any stage in a product's life cycle, including production and during the disposal of the unused drugs, containers, and waste material containing the product (manure, fish water, and other dirty water) (Montforts 1999). A summary of the possible emission routes to the environment is given next. The importance of individual routes into the environment for different types of medicine varies according to the type of treatment, the route of administration, and the type of animal being treated.

A. Emissions During Manufacturing and Formulation

During the manufacture of an active pharmaceutical ingredient (API) and formulation of the finished drug product, raw materials, intermediates, or the active substance may be released to the air, to water in wastewater, and to land in the form of solid waste. During manufacture, the main route of release of drugs into the environment is probably via process waste effluents produced during the cleaning of active pharmaceutical ingredient and manufacturing equipment used for coating, blending, tablet compressing, and packing (Velagaleti and Gill 2003). Biological and chemical degradation processes such as biotransformation, mineralization, hydrolysis, and photolysis are thought to remove most drug residues before process waste effluents or sludge solids are discharged to surface waters, to sewage treatment works, or released to land (Velagaleti and Gill 2003). Manufacturing plants employ a number of treatment methodologies and technologies to control and treat emissions and minimize the amount of waste produced, such as the use of condensers, scrubbers, adsorbent filters, and combustion or incineration for recovery and removal in air emissions. Neutralization, equalization, activated sludge, primary clarification, multimedia filtration, activated carbon, chemical oxidation, and advanced biological systems may be used for treatment of waste waters (USEPA 1997). In addition, a number of practices are often implemented by the industry to reduce waste generation and material losses, including process optimization, production scheduling, materials tracking, and waste stream segregation (USEPA 1997). Losses to the environment arising during the manufacture or formulation of veterinary medicine products are likely to be minimal.

B. Aquaculture

Chemotherapeutic pharmaceuticals used in fish farming are limited to antiinfective agents for parasitic and microbial diseases, anesthetic agents, and medical disinfectants. Drugs are commonly administered as medicated feed, by injection, or in the case of topical applications as a bath formulation. Bacterial infections in fish are usually treated using medicated food pellets, which are added directly to pens or cages (Hektoen et al. 1995; Samuelsen et al. 1992a).

When infected, cultured fish show reduced appetite and thus feed intake.

Consequently, a large proportion of medicated feed is not eaten, passes through the cages, and is available for distribution to other compartments. Furthermore, the bioavailability of many antibacterial agents is relatively low and drugs may also enter the environment via feces and urine (Björklund and Bylund 1991; Hustvedt et al. 1991). In recent years, improved husbandry practices have reduced the amount of waste feed generated, and more recently authorized medicines have greater bioavailability ($F > 95\%$). Nevertheless, deposition of drugs from uneaten feed or feces on or in undercage sediment can be a major route of environmental contamination for pharmaceuticals used in aquaculture (Björklund et al. 1991; Jacobsen and Berglind 1988; Lunestad 1992). Once present on or in sediment, compounds may also leach back into the water column. During periods of treatment, some of the drugs entering the environment in waste feed and feces are also taken up by exploitative wild fish, shellfish, and crustaceans (Capone et al. 1996; Björklund et al. 1990; Ervik et al. 1994; Samuelsen et al. 1992a).

Where topical applications of chemotherapeutants are made, fish are usually crowded into a small water volume for treatment (Burka et al. 1997; Grave et al. 1991). Concentrated drugs are added directly to the water of open net-pens or ponds, net-pens enclosed by a tarpaulin or tanks. Waste effluent is then either released into the surrounding water column or subject to local wastewater treatment and recycling (filters, settlement basins, and ponds) (Burka et al. 1997; Grave et al. 1991; Montforts 1999). In addition, sludge recovered from wastewater recycling activities may be applied directly to land or sold as fertilizer (Montforts 1999).

C. Agriculture (Livestock Production)

Large quantities of animal health products are used in agriculture to improve animal care and increase production. Some drugs used in livestock production are poorly absorbed by the gut, and the parent compound or metabolites are known to be excreted in the feces or urine, irrespective of the method of application (Campbell et al. 1983; Chui et al. 1990; Donoho 1987; Magnussen et al. 1991; Sommer et al. 1992; Stout et al. 1991). During livestock production, veterinary drugs enter the environment through removal and subsequent disposal of waste material, including manure/slurry and "dirty" waters, via excretion of feces and urine by grazing animals, through spillage during external application, or by direct exposure/discharge to the environment.

With all hormones, antibiotics, and other pharmaceutical agents administered either orally or by injection to animals, the major route of entry of the product into the environment is probably via excretion following use and the subsequent disposal of contaminated manure onto land (Halling-Sørensen et al. 2001). Many intensively reared farm animals are housed indoors for long periods at a time. Consequently, large quantities of farmyard manure, slurry, or litter are produced that are then disposed of at relatively high application rates onto land (ADAS 1998, 1997; Montforts 1999). Although each class of livestock production has different housing and manure production characteristics, the emission and distribution routes for veterinary medicines are essentially similar. As well as contaminating the soil column, it is possible for veterinary medicines to leach to shallow groundwater from manured fields or even reach surface water bodies through surface runoff (Hamscher et al. 2000a–c; Hirsch et al. 1999; Meyer et al. 2000; Nessel et al. 1989). In addition, drugs administered to grazing animals or animals reared intensively outdoors are deposited directly to land or surface water in dung or urine, potentially exposing soil and/or aquatic organisms to high local concentrations (Halling-Sørensen et al. 1998; McCracken 1993; Montforts 1999; Sommer and Overgaard Nielsen 1992; Sommer et al. 1993; Strong 1992, 1993; Strong and Wall 1994).

Release of substances used in topical applications may also occur. Various substances are used externally on animals and poultry for the treatment of external or internal parasites and infection. Sheep in particular suffer from a number of external insect parasites for which treatment and protection are sometimes obligatory. The main methods of external treatment include plunge dipping, pour-on formulations, and the use of showers or jetters. With all externally applied veterinary medicines, both diffuse and point-source pollution can occur. Sheep-dipping activities provide several routes for environmental contamination. In dipping practice, chemicals may enter watercourses through inappropriate disposal of used dip, leakage of used dip from dipping installations, and from excess dip draining from treated animals. Current disposal practices rely heavily on spreading used dip onto land (HSE 1997; MAFF 1998).

Washoff of chemicals from the fleeces of recently treated animals to soil, water, and hard surfaces may occur on the farm, during transport, or at stock markets. Some market authorities insist animals are dipped before entering the market to restrict the spread of disease, thus creating the potential for contaminated runoff from uncovered standing areas (Armstrong and Philips 1998). Sheep dip chemicals may also be released to the environment from wool washing plants and fellmongers (initial processing stage of leather production) (Armstrong and Philips 1998). Monitoring data (Environment Agency 1998, 2001) have demonstrated high numbers of Environmental Quality Standard (EQS) failures in the Yorkshire area associated with the textile industry. Although effluent produced from the wool washing process is normally treated for the removal of pollutants, this process is not always adequately effective and chemicals may be released in discharges from the treatment plants. In addition, spills and leaks of untreated effluent directly to surface water drains from both fellmongers and wool treatment plants can occur (Environment Agency 1999).

Other topically applied veterinary medicines likely to wash off following use include udder disinfectants from dairy units and endectocides for treating cattle parasites. Udder washings containing antiinfective agents and contaminated dirty water produced by dairy units may enter the environment through soakaways, surface water drains, or via inclusion in stored slurry and subsequent application to land. Washoff from the coats/skin of cattle treated with pour-on formulations can occur where the animals are exposed to rain shortly after dosing (Bloom and Matheson 1993). Residues of drugs in washoff may accumulate in localized high concentrations on land with high stocking densities. Contaminated surface runoff from open cattle yards (dirty water) is often collected and subsequently spread onto land. In addition, residues may wash off the backs and coats of grazing animals such as cattle and sheep that have access to surface water bodies as drinking water.

D. Companion/Domestic Animals

To date, the environmental fate of veterinary medicines used in companion animals (pets) has not been extensively researched, probably because unlike production animals reared in agriculture, companion animals are generally kept on a small-scale basis and are therefore not subject to mass medication. Where used, drugs are likely to be dispersed into the environment via runoff or leaching from on-ground fecal material (Daughton and Ternes 1999). In addition, ectoparasiticides applied externally to canine species may contaminate surface water through direct loss from the coat if the animal enters the water.

E. Disposal of Unwanted Drugs

Veterinary pharmaceutical drugs may be subject to disposal at any stage during their life cycle. It is probably fair to assume that, as with human pharmaceuticals, a proportion of all prescribed or nonprescribed veterinary medicines will be unused and unwanted by the end user. The principal end users of veterinary medicines are veterinarians, livestock producers, or domestic users. Disposal of veterinary medicines by end users should be interpreted to include damaged, outdated, or outmoded animal medicines, as well as used containers and packages, contaminated sharps, applicators, and protective clothing (Cook 1995). Users are advised to always follow advice on the label regarding disposal and never to dispose of such items with domestic rubbish or down the drain or toilet.

Where appropriate, product label and safety data sheets provided by manufacturers give information relating to the safe disposal of veterinary medicines and packaging. In the UK, distributors, veterinary practices, farmers, and feed compounders can also contact the manufacturer or local authority for advice, especially if large quantities of animal medicines require disposal and collection services are operated by some county councils for the periodic disposal of special waste (Cook 1995). Users of companion animal products may return unwanted or unused product to the veterinary surgery or local pharmacist.

In practice, methods for disposal include flushing down the toilet, incineration, and local domestic waste collection. Domestic users will undoubtedly flush unwanted medicines down toilets or place them with the domestic refuse (Daughton and Ternes 1999). For ectoparasiticides, in particular sheep dips, containers should be returned to suppliers for correct disposal to high-temperature incineration or a licensed landfill. In the UK, if on-farm disposal is planned, water-soluble preparation containers should be triple rinsed before burning or burial away from water courses or any land drains, as specified by the Code of Good

Agricultural Practice for the Protection of Water (MAFF, 1998). Inappropriate disposal of empty containers and unwanted product by careless operators may lead to contamination of soil and waters.

Unwanted or expired products that are returned to the manufacturer are usually disposed of through incineration or landfilling at suitable sites (Velagaleti and Gill 2003). Where drugs are disposed of in sufficient quantities to unlined landfill sites, residues present in the leachate may reach shallow groundwater and surface waters (Holm et al. 1995).

V. Occurrence in the Environment

Veterinary medicines have been measured in surface waters, groundwaters, sediments, slurry/manure, and biota. Monitoring studies have focused on veterinary products used in sheep dips, in aquaculture, and as antibiotic treatments for livestock (Table 3).

A. Aquaculture

During the past two decades, a number of studies have investigated the environmental impact of chemotherapeutic drugs used in aquaculture. Antibacterial drugs are mostly given as medicated food pellets. It is well documented that the majority of orally administered chemotherapeutics ultimately leave the treated cages/lagoons as surplus food and enter the environment (Lunestad 1992; Samuelsen et al. 1992b; Thorpe et al. 1990 cited in Capone et al. 1996). To ensure cost-effective treatment, aquaculture facilities endeavor to ensure that most of an administered medicine is taken up by the target stock. A discussion of monitoring studies that have been conducted, including measured environmental concentrations, is presented next.

Emamectin Benzoate and Its Major Metabolite (4"-Epiaminoavermectin B_{1a})
Emamectin benzoate is a premix therapeutic agent, effective against several life stages of sealice. As part of an environmental risk assessment of emamectin benzoate, field monitoring studies were carried out at a fish farm sited on a Scottish loch to determine chemical residues in sediment, flocculent material retrieved from the loch bed, water, particulate matter, and deployed and indigenous fauna (SEPA 1999). Most samples collected and analyzed contained no measurable concentrations of either the parent compound or its major desmethylamino metabolite [limit of detection (LOD) water, 0.2 µg L^{-1}; LOD in sediment, flocculent material, particulate matter, and deployed and indigenous fauna, 0.25 µg kg^{-1}]. However, a maximum concentration of 5.0 µg kg^{-1} emamectin benzoate was recorded 1 week post treatment in hermit crabs, and of 1.23 and 1.99 µg kg^{-1} in dogfish and the crab species *Munida rugosa*, respectively, at the same time interval. The compound was also detected in sediment samples collected up to 12 months following treatment at levels up to 2.73 µg kg^{-1} and just above the LOD (0.25 µg kg^{-1}) in 5 of 61 samples of flocculent material collected and

Table 3. Environmental monitoring data.

Compound	Therapeutic use	Concentration detected (ng L^{-1} unless otherwise stated)	Limit of detection (LOD) (ng L^{-1} unless otherwise stated)	Country	Reference
Surface water:					
Freshwater/marine water					
Chlorfenvinphos	Ectoparasiticide	ND–30,800	5.0–400	England & Wales	EA (1997)
		<20–3,068	20	Scotland	Littlejohn and Melvin (1991)
		1–355	2–20,000	England & Wales	EA (2001)
Chloramphenicol	Antibiotic	0.06 µg L^{-1}	0.02 µg L^{-1} (LOQ)	Germany	Hirsch et al. (1999)
Chlortetracycline	Antibiotic	0.5 µg L^{-1}	0.5 µg L^{-1}	USA	Meyer et al. (2000)
	Antibiotic	ND–690	50–100	USA	Kolpin et al. (2002)
Doxycycline	Antibiotic	ND	100	USA	Kolpin et al. (2002)
Enrofloxacin	Antibiotic	ND	20	USA	Kolpin et al. (2002)
Erythromycin-H$_2$O	Antibiotic metabolite	ND–1,700	50	USA	Kolpin et al. (2002)
Lincomycin	Antibiotic	ND–730	50	USA	Kolpin et al. (2002)
Oxytetracycline	Antibiotic	ND–340	100	USA	Kolpin et al. (2002)
Sarafloxicin	Antibiotic	ND	20	USA	Kolpin et al. (2002)
Sulfachloropyridazine	Antibiotic	ND	50	USA	Kolpin et al. (2002)
Sulfadimethoxine	Antibiotic	ND–60	50	USA	Kolpin et al. (2002)
Tetracycline	Antibiotic	ND–110	50–100	USA	Kolpin et al. (2002)
Trimethoprim	Antibiotic	ND–710	14–30	USA	Kolpin et al. (2002)
Tylosin	Antibiotic	ND–280	50	USA	Kolpin et al. (2002)
Coumaphos	Ectoparasiticide	30	10–50	England & Wales	EA (1997)

Table 3. (*Continued*).

Compound	Therapeutic use	Concentration detected (ng L^{-1} unless otherwise stated)	Limit of detection (LOD) (ng L^{-1} unless otherwise stated)	Country	Reference
Cypermethrin	Ectoparasiticide	ND–200	0.1–500	England & Wales	EA (1997)
		1–85,100	1–2,000	England & Wales	EA (2001)
Desmethylamino metabolite	Derivative of emamectin benzoate	2.4 μg L^{-1}	0.2 μg L^{-1}	Scotland	SEPA (1999)
Diazinion	Ectoparasiticide	ND–2,500	1.0–1,000	England & Wales	EA (1997)
		<10–108	10	Scotland	Littlejohn and Melvin (1991)
		3–0.58 × 10^6	3	Scotland	Virtue and Clayton (1997)
		1–5,000	1–12,500	England & Wales	EA (2001)
Emamectin benzoate	Ectoparasiticide	1.06 μg L^{-1}	0.2 μg L^{-1}	Scotland	SEPA (1999)
Fenchlorphos	Ectoparasiticide	<10–777	10	Scotland	Littlejohn and Melvin (1991)
Flumethrin	Ectoparasiticide	1–2,190	1–45,000	England & Wales	EA (2001)
Propetamphos	Ectoparasiticide	ND–1,200	4.5–200	England & Wales	EA (1997)
		<10–2,173	10	Scotland	Littlejohn and Melvin (1991)
		3–19.2 × 10^6	3	Scotland	Virtue and Clayton (1997)
		1–11,738,000	1–10,000	England & Wales	EA (2001)

Table 3. (*Continued*).

Compound	Therapeutic use	Concentration detected (ng L^{-1} unless otherwise stated)	Limit of detection (LOD) (ng L^{-1} unless otherwise stated)	Country	Reference
Groundwater					
Chlorfenvinphos	Ectoparasiticide	ND–70	5.0–200	England & Wales	EA (1997)
		15–20	1–20,000	England & Wales	EA (2001)
Chlortetracycline	Antibiotic	0.17–0.22 µg L^{-1}	0.1–0.3 µg L^{-1}	Germany	Hamscher et al. (2000a)
Diazinon	Ectoparasiticide	ND–216	5.0–200	England & Wales	EA (1997)
		26–190	5–10,000	England & Wales	EA (2001)
Oxytetracycline	Antibiotic	0.15–0.19 µg L^{-1}	0.1–0.3 µg L^{-1}	Germany	Hamscher et al. (2000a)
Propetamphos	Ectoparasiticide	ND–489	4.7–100	England & Wales	EA (1997)
		29–110	1–10,000	England & Wales	EA (2001)
Sulfamethazine	Antibiotic	0.08–0.16 µg L^{-1}	0.02 µg L^{-1} (LOQ)	Germany	Hirsch et al. (1999)
Tetracycline	Antibiotic	0.11–0.27 µg L^{-1}	0.1–0.3 µg L^{-1}	Germany	Hamscher et al. (2000a)
Tylosin	Antimicrobial	0.13–0.42 ± 0.47 µg L^{-1}	0.1–0.3 µg L^{-1}	Germany	Hamscher et al. (2000a)
Surface/subsurface runoff					
Ivermectin	Endectocide	<1.1–4.4	—	USA	Nessel et al. (1989)
Sediment					
Oxytetracycline	Antibiotic	0.1–4.9 mg kg^{-1} dry matter	—	Norway	Jacobsen and Berglind (1988)
		0.05–16 µg g^{-1}	0.05 µg g^{-1}	Finland	Björklund et al. (1990)

Table 3. (*Continued*).

Compound	Therapeutic use	Concentration detected (ng L^{-1} unless otherwise stated)	Limit of detection (LOD) (ng L^{-1} unless otherwise stated)	Country	Reference
		0.8–6.3 µg g^{-1}	0.05 µg g^{-1}	Finland	Björklund et al. (1991)
		189–285 µg g^{-1}	0.1 µg g^{-1}	Norway	Samuelsen et al. (1992b)
		<1.2 µg g^{-1}–10.9 ± 6.5 µg g^{-1}	1.2 µg g^{-1}	Ireland	Coyne et al. (1994)
		0.2–4 µg g^{-1}	0.2 µg g^{-1}	USA	Capone et al. (1996)
		1.3–4.5 µg g^{-1}	1.2 µg g^{-1}	Ireland	Kerry et al. (1996)
Desmethylamino metabolite	Derivative of emamectin benzoate	>0.5 µg kg^{-1} (wet wt)	0.25 µg kg^{-1} (wet wt)	Scotland	SEPA (1999)
Emamectin benzoate	Ectoparasiticide	0.25–2.73 µg kg^{-1}	0.25 µg kg^{-1} (wet wt)	Scotland	SEPA (1999)
Ivermectin	Ectoparasiticide	Trace–6.8 ng g^{-1}	0.5 ng g^{-1}	Ireland	Canavan et al. (2000)
Oxolinic acid	Antimicrobial	<0.05–0.2 µg g^{-1}	0.05 µg g^{-1}	Finland	Björklund et al. (1991)
Suspended particulate matter					
Desmethylamino metabolite	Derivative of emamectin benzoate	1.9–30 µg kg^{-1}	0.25 µg kg^{-1} (wet wt)	Scotland	SEPA (1999)
Emamectin benzoate	Ectoparasiticide	75.1–366 µg kg^{-1}	0.25 µg kg^{-1} (wet wt)	Scotland	SEPA (1999)
Flocculent material from seabed					
Emamectin benzoate	Ectoparasiticide	Trace	0.25 µg kg^{-1} (wet wt)	Scotland	SEPA (1999)

Table 3. (Continued).

Compound	Therapeutic use	Concentration detected (ng L^{-1} unless otherwise stated)	Limit of detection (LOD) (ng L^{-1} unless otherwise stated)	Country	Reference
Soil					
Chlortetracycline	Antibiotic	0.7 ± 0.2–9.5 ± 2.8 μg kg^{-1}	0.7 μg kg^{-1}	Germany	Hamscher et al. (2000a)
		<1–26.4 μg kg^{-1}	1 μg kg^{-1}	Germany	Hamscher et al. (2000b)
		1.2–41.8 μg kg^{-1}	1 μg kg^{-1}	Germany	Hamscher et al. (2000c)
Ivermectin	Endectocide	0.1–2 μg kg^{-1}	1 μg kg^{-1}	USA	Nessel et al. (1989)
Monensin	Coccidiostat	0.8–1.08 mg kg^{-1}	—	Canada	Donoho (1984)
Oxytetracycline	Antibiotic	0.9 ± 0.1–8.6 ± 4.5 μg kg^{-1}	0.7 μg kg^{-1}	Germany	Hamscher et al. (2000a)
Tetracycline	Antibiotic	1.2 ± 0.1–12.3 ± 5.6 μg kg^{-1}	0.7 μg kg^{-1}	Germany	Hamscher et al. (2000a)
		<1–32.2 μg kg^{-1}	1 μg kg^{-1}	Germany	Hamscher et al. (2000b)
		1.1–39.6 ± 33.6 μg kg^{-1}	1 μg kg^{-1}	Germany	Hamscher et al. (2000c)
Tylosin	Antimicrobial	nd/trace	0.2 μg kg^{-1}	Germany	Hamscher et al. (2000a)
Feces and urine					
Cattle feces/manure					
[^{14}C]Ceftiofur	Antibiotic	11.3–216.1 mg kg^{-1} (equivalent)	—	USA	Gilbertson et al. (1990)

Table 3. (*Continued*).

Compound	Therapeutic use	Concentration detected (ng L^{-1} unless otherwise stated)	Limit of detection (LOD) (ng L^{-1} unless otherwise stated)	Country	Reference
Chlortetracycline	Antibiotic	7.6 ± 2.7 μg kg^{-1}	—	Germany	Hamscher et al. (2000c)
Ivermectin	Endectocide	12–75 μg kg^{-1}	10 μg kg^{-1}	USA	Nessel et al. (1989)
		0.3 ± 0.0–9.0 ± 0.7 mg kg^{-1}	0.03 mg kg^{-1}	Denmark	Sommer and Steffansen (1993)
		0.2–3.8 mg kg^{-1} (dry wt)	—	Tanzania	Sommer and Overgaard Nielsen (1992)
		0.07–0.36 mg kg^{-1} (wet wt)	—	Australia	Cook et al. (1996)
		0.353 mg kg^{-1}	—	USA	Merck, Sharpe & Dohme (1983) [cited in Strong (1992)]
		13–80 μg kg^{-1}	—	USA	Halley et al. (1986)
		0.24–0.27	—	USA	Halley et al. (1989)
Monensin	Coccidiostat	0.7–4.7	—	Canada	Donoho (1984)
Sulfadimethoxine	Antimicrobial	300–900 mg kg^{-1}	—	Italy	Brambilla, unpublished data (1993) [cited in Migliore et al. (1995)]
Tetracycline	Antibiotic	2.5 ± 1.2 μg kg^{-1}	—	Germany	Hamscher et al. (2000c)

Table 3. (Continued).

Compound	Therapeutic use	Concentration detected (ng L^{-1} unless otherwise stated)	Limit of detection (LOD) (ng L^{-1} unless otherwise stated)	Country	Reference
Pig feces/manure					
Chlortetracycline	Antibiotic	3.4–1001.6 µg kg^{-1}	—	Germany	Hamscher et al. (2000c)
Ivermectin	Endectocide	0.22–0.24 mg kg^{-1}	—	USA	Halley et al. (1989)
Tetracycline	Antibiotic	44.4–132.4 µg kg^{-1}	—	Germany	Hamscher et al. (2000c)
Sheep feces/manure					
Ivermectin	Endectocide	0.63–0.714 mg kg^{-1}	—	USA	Halley et al. (1989)
Poultry feces/manure					
Chlortetracycline	Antibiotic	22.5 µg g^{-1}	—	Canada	Warman and Thomas (1981)
[^{14}C]Narasin	Antibiotic	1 ± 0.3–725 ± 60.3 µg kg^{-1}(equivalent)	—	USA	Catherman et al. (1991)
Horse feces/manure					
Ivermectin	Endectocide	0.05–8.47 µg g^{-1}	0.05 µg g^{-1}	USA	Jernigan et al. (1990) [cited in Sams (1993)]
Fauna					
Wild fish					
Emamectin benzoate	Ectoparasiticide	0.25–1.23 µg kg^{-1}	0.25 µg kg^{-1} (wet wt)	Scotland	SEPA (1999)
Oxytetracycline	Antibiotic	0.05–1.3 µg g^{-1}	0.05 µg g^{-1}	Finland	Björklund et al. (1990)

Table 3. (Continued).

Compound	Therapeutic use	Concentration detected (ng L^{-1} unless otherwise stated)	Limit of detection (LOD) (ng L^{-1} unless otherwise stated)	Country	Reference
Oxolinic acid	Antimicrobial	0.01–13.59 µg g^{-1}	0.003–0.01 µg g^{-1}	Norway	Samuelsen et al. (1992a)
Flumequine	Antimicrobial	<0.08–15.74 µg g^{-1} 0.06–1.12 µg g^{-1} (mean conc)	— —	Norway Norway	Ervik et al. (1994) Ervik et al. (1994)
Shellfish					
Desmethylamino metabolite	derivative of emamectin benzoate	Trace	0.25 µg kg^{-1} (wet wt)	Scotland	SEPA (1999)
Emamectin benzoate	Ectoparasiticide	Trace	0.25 µg kg^{-1} (wet wt)	Scotland	SEPA (1999)
Oxolinic acid	Antimicrobial	0.03–3.77 µg g^{-1}	0.003–0.01 µg g^{-1}	Norway	Samuelsen et al. (1992a)
Crustacea					
Emamectin benzoate	Ectoparasiticide	0.25–5 µg kg^{-1}	0.25 µg kg^{-1} (wet wt)	Scotland	SEPA (1999)
Oxolinic acid	Antimicrobial	0.05–1.48 µg g^{-1}	0.003–0.01 µg g^{-1}	Norway	Samuelsen et al. (1992a)
Oxytetracycline	Antibiotic	0.1–3.8 µg g^{-1}	0.1 µg g^{-1}	USA	Capone et al. (1996)

ND: not determined

analyzed. Water and particulate components collected from silt traps suspended 2 m above the loch bed were analyzed separately. The parent compound was detected at 1.06 µg L^{-1} in water and at 75.1, 154, and 366 µg kg^{-1} in the particulate component. In the same study, the desmethylamino metabolite was detected infrequently in sediment and mussel samples above the limit of quantification (0.5 and 1.0 µg kg^{-1}, respectively). A peak concentration of 30 µg kg^{-1} of the metabolite was detected in the particulate component of samples collected in silt traps and at a concentration of 2.4 µg L^{-1} in the water component of a pretreatment silt trap sample.

Oxolinic Acid Studies have shown residues of oxolinic acid to be present in the surrounding wild fish population and other marine animals during and after the medication of cultivated fish (Ervik et al. 1994; Samuelsen et al. 1992a). In both studies, wild fauna were captured and monitored within the vicinity of aquaculture facilities off the west coast of Norway following treatment with oxolinic acid. Maximum reported concentrations of oxolinic acid in the two studies were 15.74 µg g^{-1} for fish muscle, 3.77 µg g^{-1} for crab muscle, and 1.48 µg g^{-1} for mussel tissue. Samuelsen et al. (1992a) demonstrated that tissue concentrations had declined to low levels 12 d after treatment.

In a study conducted off the southwest coast of Finland (Björkland et al. 1991), residues of oxolinic acid were detected in anoxic sediments collected below three of five fish farms where fish had been treated. Maximum concentrations of 0.05–0.2 µg g^{-1} were measured in sediments for 5 d after treatment of the fish.

Oxytetracycline The environmental fate of oxytetracycline following its use in aquaculture has been extensively researched (Björklund et al. 1990, 1991; Capone et al. 1996; Coyne et al. 1994; Jacobsen and Berglind 1988; Kerry et al. 1996; Samuelsen et al. 1992a).

A limited number of studies have investigated residues of oxytetracycline in wild fauna (Björklund et al. 1990; Capone et al. 1996). Concentrations of oxytetracycline in samples of bleak and roach obtained from around a Finnish fish farm on the last day of medication ranged from 0.06 to 3 µg g^{-1} (bleak) and from 0.05 to 0.1 µg g^{-1} (roach) in the muscle tissue (Björklund et al. 1990). In bleak, concentrations declined to levels at or near the limit of detection soon after treatment had finished, whereas in the roach measurable concentrations were observed in some fish samples up to 13 d after treatment. Similar low concentrations of oxytetracycline in wild fauna were also observed in a study conducted in Puget Sound, Washington (USA) (Capone et al. 1996), only trace oxytetracycline residues (about 0.1 µg g^{-1}) being found in oysters or Dungeness crab. However, the authors reported drug residues between 0.8 and 3.8 µg g^{-1} in the edible crabmeat of red rock crabs up to 12 d after treatment. Trace concentrations were detected in two red rock crabs collected at 41 and 75 d.

There is considerable evidence to show that the enriched sediments often present under fish farm cages contain residues of oxytetracycline (Björklund et al. 1990, 1991; Capone et al. 1996; Coyne et al. 1994; Jacobsen and Berglind 1988; Kerry et al. 1996; Samuelsen et al. 1992a). Rapid sedimentation is a process characteristic of many aquaculture facilities because of debris, mainly feces and uneaten food, leaving the cages and accumulating underneath them. Consequently, sediments containing oxytetracycline may be quickly buried and the drug may persist indefinitely. In Norway, oxytetracycline has been found at concentrations ranging from 0.1 to 4.9 mg kg^{-1} dry matter (Jacobsen and Berglind 1988). The authors indicate that antimicrobial effects might be expected at these concentrations.

In a study located in the Baltic Sea, sediment samples collected from two farms on the last day of medication were shown to contain oxytetracycline at concentrations ranging from 0.05 to 3.8 µg g^{-1} (Björklund et al. 1990). Eight days after medication had ceased, drug levels at one farm had decreased to below the detection limit (0.05 µg g^{-1}). However, up to 16 µg g^{-1} was measured in sediments taken at the other farm on day 8, and at 308 d the bottom deposits still contained between 1.0 and 4.4 µg g^{-1} sediment. The authors indicate that lower temperature and stagnant, anoxic conditions were probably responsible for the high concentrations observed.

In a separate study conducted off the southwest coast of Finland, five separate fish farms were monitored during and up to 12 d after treatment (Björklund et al. 1991). The maximum concentrations of oxytetracycline detected in the sediments were between 2.0 and 6.3 µg g^{-1}. Twelve days after the end of medication, levels of the drug had decreased to between 0.8 and 2.5 µg g^{-1}.

Similarly, low concentrations are reported in an investigation conducted at a marine salmon farm situated in Galway Bay, Ireland (Coyne et al. 1994). Oxytetracycline was detected in the top 2 cm of sediment samples collected from under two adjacent cage blocks following the therapeutic use of the drug. Peak concentrations of 10.9 ± 6.5 µg g^{-1} and 9.9 ± 2.9 µg g^{-1} were detected on the 10th d of treatment and 3 d after its last use, from under cage blocks 6 and 7, respectively. Approximately 1 mon after treatment, mean concentrations had decreased to between 1.6 ± 0.4 and 2.3 ± 0.5 µg g^{-1}. At 66 and 71 d after the end of therapy, concentrations were below the limit of detection.

In a later cage block study at the same site in Galway Bay, oxytetracycline was detected at concentrations ranging from 1.3 to 4.5 µg g^{-1} in the top 2 cm of 4 of 11 sediment cores collected 5 d after the last administration of medicated feed (Kerry et al. 1996). The authors noted that the lower concentrations are probably as a result of the reduced treatment rate; 20 kg of the agent was used in this study as opposed to the 175 kg used previously.

Capone et al. (1996) presented an extensive study consisting of field investigations at three salmon mariculture facilities in Puget Sound, Washington (USA). The farms studied were chosen to represent a gradient in the magnitude of antibacterial usage. The frequency of detection of oxytetracycline was shown to parallel drug use. Residues were rarely detected beneath a farm that used very little oxytetracycline (3 kg); however, concentrations between 0.5 and 4 µg g^{-1} were commonly detected at a farm that used 186 kg in a single prophylactic treatment period. Significantly, oxytetracycline residues (0.2–2 µg g^{-1}) were measured in surficial and subsurface sediments before treatment. The authors believe that these persistent residues are probably due to drug usage during the previous summer or earlier.

In contrast to the foregoing investigations, much larger concentrations of oxytetracycline were detected by Norwegian researchers under a salmon farm situated off the west coast of Norway (Samuelsen et al. 1992b). Following a single 10-d therapeutic use of the drug, peak concentrations of 189 and 285 µg g^{-1} were detected in under-cage sediment cores collected over a period of 18 mon after medication. The disparity in results obtained in this study with previous studies is considered an artefact of gross overfeeding at the farm (Coyne et al. 1994; Kerry et al. 1996; Smith 1996).

Flumequine In Europe, flumequine is only permitted for use in Norway (Alderman and Hastings 1998). To date, only a single study has sought to quantify environmental concentrations following use of the compound in aquaculture. Researchers in Norway recorded mean muscle concentrations between 0.06 and 1.12 mg kg^{-1} in the muscle of wild fish caught in the vicinity of a farm the day after termination of treatment (Ervik et al. 1994).

Ivermectin Following oral administration, ivermectin is mainly excreted in an unchanged form (Høy et al. 1990). Given this, a variety of modeling approaches have attempted to estimate the extent to which orally administered ivermectin will accumulate in sediments under fish farms (Davies et al. 1998). The presence of ivermectin in sediments has also been investigated at a small number of commercial fish farms. Unpublished work from two studies, for which a limit of quantitation of 10 and 50 ng g^{-1} was achieved, failed to detect any ivermectin residues in sediments (Kwok, unpublished data; Nixon E, unpublished data, cited in Canavan et al. 2000). In a third monitoring study, quantifiable residues of ivermectin, measured as H_2B_{1a}, the secondary butyl compound of ivermectin, were detected in sediments under and adjacent to salmon cages situated approximately 1 km offshore on the west coast of Ireland (Canavan et al. 2000). Sediment cores were collected on the final day of a 4-mon period in which the drug was administered twice weekly. Ivermectin was detected at concentrations between 1.4 and 6.8 ng g^{-1} to a depth of 12 cm in cores collected from under cages and up to 31 m away from the edge of the cage block. In addition, analysis of the top 2 cm of three sediment samples that had previously been collected from the same farm but stored for 4–5 yrs revealed H_2B_{1a} concentrations between 1.4 and 5.6 ng g^{-1}.

B. Agriculture

Sheep-Dipping Chemicals Extensive monitoring studies have been performed in the UK for sheep dip chemicals (Environment Agency 1998, 2001). Monitoring data for England and Wales in the year 2000 demonstrated that the organophosphate substances diazinon and propetamphos were detected in surface waters more frequently than chlorfenvinphos and the pyrethroids. Diazinon was detected in 498 of 4186 samples at concentrations ranging from 1 to 550 ng L^{-1}, whereas propetamphos was detected in 168 of 3773 samples at concentrations ranging from 1 to 11,738,000 ng L^{-1}. Chlorfenvinphos, cypermethrin, and flumethrin were detected in much fewer samples at 1–242 ng L^{-1}, 1–85,100 ng L^{-1}, and 1–2,190 ng L^{-1} respectively. Chlorfenvinphos, diazinon, and propetamphos were detected infrequently in groundwaters and marine waters, with maximum reported concentrations being 20, 240, and 58 ng L^{-1}, respectively.

Antibacterials and Anthelmintics in Soil Several veterinary drugs have been detected in soil that has been amended with animal manure. To date, the majority of data have been produced by a group of researchers in Germany. In three separate investigations, soil samples collected from regions with intensive livestock production were analyzed for frequently used drugs (Hamscher et al. 2000a–c). In the first study, soil samples were collected at various depths from eight fields in the Lower Saxony region that had been manured with slurry 2 d before sampling (Hamscher et al. 2000a). In the upper 10 cm of the soil samples, 9–12 µg kg^{-1} of chlortetracycline, oxytetracycline, and tetracycline were detected, with trace concentrations of tylosin also found. Concentrations of the three tetracycline compounds decreased with depth to about 1 µg kg^{-1} below 60 cm.

In a subsequent study conducted in Northern Germany, soil samples were collected and analyzed from 12 different agricultural fields, 4–5 mon after being treated with animal slurry (Hamscher et al. 2000b). Tetracycline and chlortetracycline were detected in the top 30 cm of nearly all samples at concentrations between 1–32.2 and 1.2–26.4 µg kg^{-1}, respectively. In a follow-on study conducted by the same researchers, the average distribution of tetracycline in the top 30 cm of soil amended with animal slurry was between 20 and 40 µg kg^{-1} (Hamscher et al. 2000c). Levels of chlortetracycline were generally below 5 µg kg^{-1}, although a peak concentration of 41.8 µg kg^{-1} was detected at a depth of 0–10 cm in one soil sample.

Elsewhere, information on residues of veterinary medicines in soil is particularly scarce. American researchers detected trace amounts (~0.1–2 µg kg^{-1}) of ivermectin in the top (0–7.5 cm) of soil in a cattle feedlot housing animals treated 28 d previously (200 µg kg^{-1} body weight) (Nessel et al. 1989). The authors suggested the concentrations detected in the soil are probably as a result of the feces being trampled into the mud and subsequently being protected from light, thus retarding degradation.

Surface Water Whilst monitoring sewage treatment screens and associated surface waters for 18 different antibiotic substances, residues of chloramphenicol were detected by German researchers at 0.06 and 0.56 µg L^{-1} (Hirsch et al. 1999). The authors point out that as its use in human medicine is extremely limited, the two positive detections are more likely to result from its sporadic veterinary use in fattening farms.

In studies for the Centers for Disease Control and Prevention (CDC), the U.S. Environmental Protection Agency (USEPA) and U.S. Geological Survey (USGS) sampled and analyzed liquid waste from hog lagoons, 13 in three states, and surface and groundwater from areas associated with intensive swine and poultry production, 52 from seven states (Meyer et al. 2000). All samples were analyzed for chlortetracycline. Although the compound was detected at up to several hundred parts per billion in lagoon samples, it was only found in one surface water sample, at 0.5 µg L^{-1} (LOD).

A number of veterinary medicines (including chlortetracycline, erythromycins-H_2O, lincomycin, oxytetracycline, sulfadimethoxine, tetracycline, trimethoprim, and tylosin) were detected in a recent monitoring study in the U.S. (Kolpin et al., 2002). The maximum reported concentration of 1.7 mg L^{-1} was observed for erythromycin-H_2O.

Groundwater There are only a few reports of veterinary medicines being detected in groundwater (Hamscher et al. 2000a; Hirsch et al. 1999). In an extensive monitoring study conducted in Germany, a large number of groundwater samples were collected from agricultural areas to determine the extent of contamination by antibiotics (Hirsch et al. 1999). The data show that in most areas with intensive livestock breeding, no antibiotics were present above the LOD (0.02–0.05 µg L^{-1}). Sulfonamide residues were, however, detected in four samples. Although the source of contamination of two of these is considered to be attributable to irrigation with sewage, the authors conclude that sulfamethazine, detected at concentrations of 0.08 and 0.16 µg L^{-1}, could possibly have derived from veterinary applications, as it is not used in human medicine.

In the investigations of Hamscher et al. (2000a), soil water was collected and analyzed from four separate areas of agricultural land: two belonging to livestock farms and treated with animal slurry, and two where no animal manure had been applied for approximately 5 yr. Chlortetracycline, oxtetracycline, tetracycline, and tylosin were all found at the LOD (0.1–0.3 µg L^{-1}) in water samples collected at 80 and 120 cm depth, independent of soil treatment. In addition, no biologically active residues could be detected with microbiological assays that had approximately fivefold-higher detection limits.

Veterinary medicines are also known to leach from landfill sites. In Denmark, high concentrations (parts per million) of numerous sulfonamides were found in leachates close to a landfill site where a pharmaceutical manufacturer had previously disposed of large amounts of these drugs over a 45-yr period (Holm et al. 1995). Concentrations dropped off significantly tens of meters down gradient, most probably a result of microbial attenuation. Although this is recognized as a specific problem, the disposal of smaller quantities of veterinary medicines to a landfill should nevertheless be considered a potential route for environmental contamination.

Surface/Subsurface Runoff So far, only two studies have investigated the occurrence of veterinary drugs in surface or subsurface runoff. In a post-approval study carried out for Merck & Co., the runoff from a cattle feedlot following injection of five steers with ivermectin at 200 µg kg^{-1} body weight was collected and analyzed for six separate time periods (Nessel et al. 1989). Samples were collected during the 7 d before treatment, to establish baseline data, and during four consecutive 7-d periods following injection. The authors reported trace amounts of ivermectin (1.1–1.2 ng L^{-1}) detected in two surface water samples collected, 0–6 and 14–20 d post treatment, and 2 ng L^{-1} of ivermectin in the surface water of a pen flood irrigated on day 28 after the treated animals had been removed. In the 7-d period before treatment, ivermectin was detected at 3.2–4.4 ng L^{-1} and 0.8–1.5 ng L^{-1} in surface and subsurface water, respectively.

A recent study by Boxall et al. (2002) demonstrated that antibiotics applied to soil in pig slurry are rapidly transported to field drains and are then transported to surface waters in subsurface drainage systems. Maximum concentrations in drainflow of 590 µg L^{-1} were recorded for the sulfonamide antibiotic sulfachloropyridazine.

Runoff from Topical Application Veterinary drugs in topically applied formulations have the potential to be washed off the backs of treated animals exposed to rain shortly after dosing. In a wash-off study conducted by Merck & Co., animals were treated with a topical dose of ivermectin (500 µg kg^{-1} body weight) and then 6 hr later subjected to 12.5-mm artificial rainfall over a 10-min period (Bloom and Matheson 1993). Approximately 0.6% (714 µg) of the applied dose was recovered in the wash-off water (5.4 L). The average concentration of ivermectin was determined to be 1.32 µg L^{-1}.

VI. Metabolism and Environmental Fate

Once administered to an animal, veterinary products may be metabolized and the resulting metabolites excreted, with any remaining parent compound, in urine and feces. The resulting excreta may be released directly to land or stored and applied to land at a later stage. Once released into the environment, veterinary medicines will be transported and distributed among the major environmental compartments (soil, air, surface waters, sediment, and biota). The resulting concentrations in these compartments will be determined by a number of factors and processes, including dosage of compound, the physicochemical properties of the substance, degradation in manure and slurry, partitioning to soil and sediment, abiotic and biotic degradation, and environmental characteristics including soil type and climatic conditions. A number of studies have investigated the metabolism and environmental behavior of veterinary medicines. The available data are given in Tables 4, 5, and 6, and a detailed discussion of some of these factors and processes follows.

A. Metabolism of Veterinary Medicines

For compounds that are administered by injection, some of the dose may remain at the injection site for some time and therefore may not be absorbed. For compounds that are administered orally, the amount absorbed can range from a small proportion to about 100%. Once absorbed, the product may undergo phase 1 metabolism followed by phase II metabolism. These reactions may produce polar metabolites that are excreted in the urine or feces. If the compound is not metabolized, then it may be excreted unchanged. Consequently, animal feces may well contain a mixture of the parent compound and metabolites. The environmental impact of the parent and major metabolites should therefore be considered in any assessment of risk. A summary of the metabolism of the major therapeutic classes of veterinary medicinal products is given in Table 4.

B. Fate in Manure and Slurry

On livestock farms where animals are housed, large quantities of farmyard manure (animal urine and feces along with fouled bedding material) and/or slurry (urine, feces, and washing-down water) are produced. Both can be either stored in manure pits for subsequent application or applied immediately to land as an organic matter supplement and fertilizer (Velagaleti and Gill 2003). The storage time for slurry varies from 0 to 50 mon, with an average of 9 mon, and for manure from 0 to 48 mon, with an average of about 6 mon (WRc-NSF 2000). Consequently, there is the potential for veterinary medicines to be degraded during a period of storage.

Data are available on the persistence in manure of a range of commonly used classes of antibiotic veterinary medicines (Table 5). Sulfonamides, aminoglycosides, β-lactams, and macrolides have half-lives of 30 d or less and are therefore

Table 4. Metabolism of major therapeutic classes of veterinary medicines.

Therapeutic class	Chemical group	Metabolism
Antimicrobials	Tetracyclines	Minimal
Antimicrobials	Potentiated sulphonamides	High
Endoparasiticides, coccidiostats	—	Moderate–high
Antimicrobials	Macrolides	Minimal
Antimicrobials	Aminoglycosides	Minimal–high
Endoparasiticides, wormers	Azoles	Moderate
Endoparasiticides, wormers	Macrolide endectins	Minimal–moderate
Antimicrobials	Others	Moderate–high
Antimicrobials	Lincosamides	Moderate
Antimicrobials	Fluoroquinolones	Minimal–high
Endoparasiticides, antiprotozoals	—	Minimal–high
Endectocides	Macrocyclic lactones	Minimal–high

Minimal: <20%; moderate: 20%–80%; high: >80%.

likely to be significantly degraded during manure/slurry storage (although no data are available on the fate of the degradation products). In contrast, ivermectin, tetracyclines, and quinolones have longer half-lives and are therefore likely to be more persistent.

C. Fate in Soil

When a veterinary medicine reaches the soil, it may partition to the soil particles, run off to surface water, leach to groundwater, or be degraded. Data are available on the sorption behavior of antibiotics, sheep dip chemicals, and avermectins in soils (Table 6). The degree to which veterinary medicines may adsorb to particulates varies widely. Consequently, the mobility of different veterinary medicinal products also varies widely. Partition coefficients (K_d) range from low (0.6 L kg^{-1}) to high (6000 L kg^{-1}) adsorption (K_{OC}, the organic normalized partition coefficient, ranges from 40 to 1.63×10^7 L kg^{-1}). In addition, the variation in partitioning for a given compound in different soils can be significant, up to a factor of 30 for efrotomycin. This variation does not appear to be reduced by normalization to the organic carbon content of the soils for most of the compounds.

The range of partitioning values can be explained by studies addressing the sorption of tetracycline and enrofloxacin. The results suggest that surface interactions of these compounds with clay minerals are responsible for the strong sorption to soils. The underlying processes are cation exchange (tetracycline at low pH) and surface complexation with divalent cations sorbed at the clay surfaces (tetracycline at intermediate pH and enrofloxacin). Thus, to arrive at a realistic assessment of the availability of these compounds for transport through the soil and uptake into soil organisms, soil chemistry may not be reduced to the organic carbon content, and the clay content, pH of the soil solution, and the coverage of the ion-exchange sites need to be considered.

The main route for degradation of veterinary medicines in soils is via aerobic soil biodegradation. Degradation rates in soil vary, with half-lives ranging from days to years (Table 7). Degradation of veterinary medicines is affected by environmental conditions such as temperature and pH and the presence of specific degrading bacteria that have developed to degrade groups of medicines (Gilbertson et al. 1990; Ingerslev and Halling-Sørensen 2001). As well as varying significantly between chemical classes, degradation rates for veterinary medicines also vary within a chemical class. For instance, of the quinolones, olaquindox can be considered to be only slightly persistent (half-life, 6–9 d) whereas danofloxacin is very persistent (half-life, 87–143 d). In addition, published data for some individual compounds show persistence varies according to soil type and conditions. In particular, diazinon was shown to be relatively impersistent (half-life, 1.7 d) in a flooded soil that had been previously treated with the compound, but was reported to be very persistent in sandy soils (half-life, 88–112 d) (reported in Lewis et al. 1993). Of the available data, coumaphos and emamectin benzoate were the most persistent compounds in soil with half-lives of 300 and 427 d, respectively, whereas tylosin and dichlorvos were the least persistent, with half-lives of 3–8 and <1 d, respectively.

When manure is combined with soil, degradation may be enhanced. Temperature has also been shown to significantly affect the rate of degradation of a compound. For example, a half-life of 91–217 d was recorded for ivermectin in a soil/feces mixture during winter weather conditions (Halley et al. 1993). In contrast, the compound was shown to degrade much more rapidly in a soil/feces mixture during the summer period, with a half-life of 7–14 d being measured (Halley et al. 1989). The timing of application of manure/slurry to land may therefore be a significant factor in determining the subsequent degradation rate of a compound. Depending on the nature of the chemical, other degradation and depletion mechanisms may occur, including soil photolysis and hydrolysis. The degradation products of both photolytic and hydrolytic degradation processes may undergo aerobic biodegradation in upper soil layers or anaerobic degradation in deeper soil layers.

D. Fate in Surface Waters

A number of studies have investigated the persistence of veterinary medicines in surface waters and freshwater and marine sediment (see Table 7). Substances may be degraded abiotically via photodegradation and/or hydrolysis or biotically by aerobic or anaerobic organisms. The quinolones, tetracyclines, ivermectin, and furazolidone are all rapidly photodegraded with half-lives ranging from <1 hr to 22 d (Davis et al. 1993; Halley et al. 1993; Lunestad et al. 1995; Oka et al. 1989). In contrast, trimethoprim, ormethoprim, and the sulfonamides are not readily photodegradable (Lunestad et al. 1995). Of the compounds studied in terms of potential to hydrolyze, ceftiofur is the only compound to be rapidly hydrolyzed, with a half-life of 8 d at pH 7 (Gilbertson et al. 1990). Propetam-

Table 5. Persistence of veterinary medicines in manure.

Chemical group	Compound	Half-life (d)	Persistence class[a]
Aminoglycosides	Unspecified[b]	30	Moderately persistent
Beta-lactams	Unspecified[b]	5	Slightly persistent
Macrolides	Tylosin	<2	Impersistent
	Unspecified[b]	21	Slightly persistent
Macrolide endectins	Ivermectin	>45	Moderately persistent
Quinolones	Unspecified[b]	100	Very persistent
Sulfonamides	Sulfachloropyridazine	<8	Slightly persistent
	Unspecified[b]	30	Moderately persistent
Tetracyclines	Unspecified[b]	100	Very persistent
Others	Amprolium	>8	Slightly persistent
	Meticlorpindol	>8	Slightly persistent
	Nicarbazin	>8	Slightly persistent

[a]Classification of persistence [taken from Hollis (1991)]: impersistent, $DT_{50} < 5$ d; slightly persistent, DT_{50} 5–21 d; moderately persistent, DT_{50} 22–60 d; very persistent, $DT_{50} > 60$ d.
[b]Halling-Sørensen et al. (unpublished).

Table 6. Sorption data for veterinary medicines.

Compound	Test matrix	K_d	K_{oc}	Reference
Avermectin B_{1a}	Clay loam soil	147 (131–161)	5,300	Gruber et al. (1990)
	Sand	17.4 (9.74–29.1)	30,000	
	Silt loam soil	80.2 (30.2–144)	6,600	
Chlorfenvinphos	—	—	295	Briggs (1981)
Ciprofloxacin	Centric flurisol	427	61,000	Nowara et al. (1997)
Efrotomycin	Silt loam soil	18	1,460	Yeager and Halley (1990)
	Loam soil	8.3	580	
	Sandy loam soil	51	8,000	
	Clay loam soil	290	11,000	
Enrofloxacin	Rhodic ferrasol	3,037	186,342	Nowara et al. (1997)
	Glegic cambisol	5,612	768,740	
	Haplic podsol	1,230	99,975	
	Rendzic leptosol	260	16,506	
	Centric flurisol	496	70,914	
	Montmorollonite	6,310	—	
	Kaolinite	3,548	—	
	Illite	4,670	—	
	Vermiculite	5,986	—	
Coumaphos	—	—	5,778–21,120	Tomlin (1997)
Deltamethrin	—	—	460,000–16,300,000	Tomlin (1997)
Diazinon	—	—	229	Briggs (1981)
	—	—	1,549	Melancom et al. (1986)
FCQA	Centric flurisol	285	40,714	Nowara et al. (1997)
Ivermectin	Clay loam	333	12,600	Halley et al. (1989)
	Silty clay loam	227	15,700	
Metronidazole	Sandy loam soil	0.67	42	Rabølle and Spliid (2000)
	Sandy soil	0.54	39	
	Sandy loam soil	0.62	56	
	Loamy sand soil	0.57	38	
Ofloxacin	Centric flurisol	309	44,143	Nowara et al. (1997)
Olaquindox	Sandy loam soil	1.67	104	Rabølle and Spliid (2000)
	Sandy soil	1.21	86	
	Sandy loam soil	1.27	116	
	Loamy sand soil	0.69	46	

Table 6. (*Continued*).

Compound	Test matrix	K_d	K_{oc}	Reference
Oxytetracycline	Sandy loam soil	680	42,506	Rabølle and Spliid (2000)
	Sandy soil	670	47,881	
	Sandy loam soil	1,026	93,317	
	Loamy sand soil	417	27,792	
Sulfamethazine	Clay loam soil	0.6	60	Thurman and Lindsey (2000)
Tetracycline	Clay loam soil	>400	40,000	Thurman and Lindsey (2000)
Tylosin	Sandy loam soil	128	7,988	Rabølle and Spliid (2000)
	Sandy soil	10.8	771	
	Sandy loam soil	62.3	5,664	
	Loamy sand soil	8.3	553	

FCQA: fluorochloroquinolone carboxylic acid.

phos was rapidly hydrolyzed at pH 3 (11 d), but hydrolysis at pH 6 and 9 was slower (1 yr and 41 d, respectively) (Lewis 1998).

Of the organophosphorous compounds that have previously been authorized for use in ectoparasitic sheep dip preparations, chlorfenvinphos, coumaphos, and dichlorvos are all relatively impersistent in biologically active water, with half-lives ranging from <1 to <25 d (Lewis 1998; Lewis et al. 1993; Tomlin 1997). Flumethrin, a synthetic pyrethroid also used as a sheep dip ectoparasiticide, was much more persistent in water, with a half-life >3 mon.

E. Fate in Sediment

A large body of data exists on the degradability of veterinary medicines used for aquaculture in both marine and freshwater sediments (Bjorklund et al. 1990; Bohm 1996; Chien et al. 1999; Coyne et al. 1994; Hansen et al. 1993; Hektoen et al. 1995; Jacobsen and Berglind 1988; Lai et al. 1995; Lunestad et al. 1995; Marengo et al. 1997; Pouliquen et al. 1992; Samuelsen 1989; Samuelsen et al. 1991, 1992b, 1994). Of the compounds studied to date, florfenicol, chloramphenicol, and furazolidone were the least persistent, with half-lives of 0.4–18.4 d (see Table 7). The other substances studied (flumequine, ormethoprim, oxytetracycline, oxolinic acid, sarafloxacin, sulfadiazine, sulfadimethoxine, and trimethoprim) persisted in sediments with half-lives being generally >30 d.

VII. Environmental Hazard

With the exception of coccidiostats, pyrimidine wormers, growth promoters, barbiturates, cephalosporin derivatives, biguanide/gluconates, NSAIDs, hormones, antiprotozoals, and enteric preparations, data were publicly available on

Table 7. Degradation data for veterinary medicines in soil, sewage sludge, and manure.

Compound	Test matrix/system	Half-life (d)	Reference
Ampicillin	Sewage treatment	48% biodegradable	Richardson and Bowron (1985)
Amprolium	Laying hen feces	30% degraded after 3 mon	van Dijk and Keukens (2000)
	Broiler feces	34% degraded after 8 d	
Bacitracin	Soil and chicken manure (20 °C)	22.5	Gavalchin and Katz (1994)
	Soil and chicken manure (30 °C)	12	
Bambermycins	Soil and chicken manure (20 °C)	<25	Gavalchin and Katz (1994)
	Soil and chicken manure (30 °C)	<30	
Ceftiofam	OECD 301 D	10% degraded after 40 d	Al-Ahmad et al. (1999)
Ceftiofur	Clay loam soil	22.2	Gilbertson et al. (1990)
	Sandy soil	49.0	
	Silty clay loam soil	41.4	
	Aqueous hydrolysis (pH 5)	100	
	Aqueous hydrolysis (pH 7)	8.0	
	Aqueous hydrolysis (pH 9)	4.2	
Chloramphenicol	Sediment (aerobic)	<12	Lai et al. (1995)
	Sediment (anaerobic)	<4	
	Marine sediment (aerobic)	2.4–18.4	Chien et al. (1999)
	Marine sediment (anaerobic)	0.4–2.4	
	Freshwater sediment (aerobic)	Rate 1.9–6.6 mg L^{-1} d^{-1}	Bohm (1996)
	Freshwater sediment (anaerobic)	Rate 20.6–24.8 mg L^{-1} d^{-1}	
	Marine sediment (aerobic)	Rate 1.9–6.0 mg L^{-1} d^{-1}	
	Marine sediment (anaerobic)	Rate 17.7–20.9 mg L^{-1} d^{-1}	
Chlorfenvinphos	Field dt_{50}	4–30 wk	Reported in Lewis (1998)
	Biodegradation in water	<25	
	Sandy loam	<5 wk	

Table 7. (*Continued*).

Compound	Test matrix/system	Half-life (d)	Reference
	High OM soil	< 9 wk	Reported in Lewis et al. (1993)
	Sand	< 4 mon	
	Peat	< 4 mon	
Chlorhexidine	Sewage treatment	Nondegradable	Richardson and Bowron (1985)
Chlortetracycline	Soil and chicken manure (30 °C)	44% removed after 30 d	Gavalchin and Katz (1994)
	Soil and chicken manure (20 °C)	No degradation after 30 d	
Ciprofloxacin	OECD 301 D	No degradation after 40 d	Al-Ahmad et al. (1999)
Coumaphos	Photolysis on soil surface	23.8 d	Tomlin (1997)
	Distilled water, pH 4	33	Reported in Lewis et al. (1993)
	Pond water, pH 5.5	<7	
	Distilled water, pH 7.0	347	
	Distilled water, pH 8.5	29	
	Sandy loam soil	300	
	Silty loam soil	200	
Cypermethrin	Hydrolysis in soil	Within 16 wk	Tomlin (1997)
	Degradation in water	5	
Danofloxacin	Three different soil types	87–143	Chen et al. (1997)
Deltamethrin	Microbial degradation in soil	Within 1–2 wk	Tomlin (1997)
	DT_{50} laboratory aerobic	21–25	
	DT_{50} laboratory anaerobic	31–36	
	Field DT_{50}	<23	
Diazinon	Water pH 3.1, 20 °C	0.5	Reported in Lewis et al. (1993)
	Water pH 7.4, 20 °C	185	
	Water pH 10.4, 20 °C	6	
	Solution, pH 5.0, 20 °C	3.8	
	Solution, pH 7.0, 20 °C	78	
	Solution, pH 9.0, 20 °C	40	
	Solutions 0–28% salinity, 10 °C	>100	
	Solutions 0–28% salinity, 20 °C	55 to >85	

Table 7. (*Continued*).

Compound	Test matrix/system	Half-life (d)	Reference
	Water pH 6.4–6.8	<30	
	Sterile soil, pH 4.7	43.8	
	Sterile sandy loam	88	
	Sterile organic soil	46	
	Soil (previously treated), pH 6.0	1.7	
	Soil (not previously treated)	9.9	
	Soil, type not given, 25 °C	11	
	Loam soil, 10 °C	21–35	
	Humic, sandy soil, 10 °C	112	
	Photodegradation in soil	12–132 hr	
Dichlorvos	Soil and water systems	<1	Tomlin (1997)
Emamectin benzoate	Hydrolysis pH 5.2–8.0	Stable over 6 wk	SEPA (1999)
	Hydrolysis pH 9	19.5 wk	
	Photolysis in solution	1.4–22.4	
	Photolysis/ biodegradation in soil	5	
Erythromycin	Sewage treatment	Nonbiodegradable	Richardson and Bowron (1985)
Ethinyl estradiol	Activated sludge STW	99.9% removal	Ternes et al. (1999)
	Biological filter STW	92% removal	
	Activated sludge STW	64% removal	
Florfenicol	Marine sediment (0–1 cm depth)	1.7	Hektoen et al. (1995)
	Marine sediment (5–7 cm depth)	7.3	
Flumequine	Marine sediment (0–1 cm depth)	60	Hektoen et al. (1995)
	Marine sediment (5–7 cm depth)	>300	
	Marine sediment	155	Hansen et al. (1993)
	Marine sediment	No degradation after 180 d	Samuelsen et al. (1994)
	Photodegradation in water	96% degraded after 9 d	Lunestad et al. (1995)
Flumethrin	Polluted water, 20 °C	>9 mon	Reported in Lewis (1998)
	Clean water, 30 °C	>3 mon	
Furazolidone	Marine sediment	0.75	Samuelsen et al. (1991)
	Photodegradation in water	92% degraded after 9 d	Lunestad et al. (1995)

Table 7. (*Continued*).

Compound	Test matrix/system	Half-life (d)	Reference
Ivermectin	Photodegradation in water	<0.5	Halley et al. (1993)
	Soil/feces mixtures (summer)	7–14	
	Soil/feces mixtures (winter)	91–217	
	Sandy loam soil	14–28	Bull et al. (1984)
	Clay soil	28–56	
	Sandy soil	56	
	Dung	Limited degradation after 45 d	Sommer et al. (1992)
Meropenem	OECD 301 D	7% degraded after 40 d	Al-Ahmad et al. (1999)
Meticlorpindol	Laying hen feces	68% degraded after 3 mon	Van Dijk and Keukens (2000)
	Broiler feces	12% degraded after 8 d	
Metronidazole	Clay soil	13.1–26.9	Ingerslev and Halling-Sørensen (2001)
	Sandy soil	9.7–14.7	Kummerer et al. (2000)
	Closed bottle test	Nondegradable	
Nicarbazin	Broiler feces	41% degraded after 8 d	Van Dijk and Keukens (2000)
Olaquindox	Clay soil	5.8–7.5	Ingerslev and Halling-Sørensen (in press)
	Sandy soil	5.9–8.8	
Ormethoprim	Marine sediment	<30	Samuelsen et al. (1994)
	Photodegradation in water	No degradation over 42 d	Lunestad et al. (1995)
Oxytetracycline	Marine sediment (0–1 cm depth)	151	Hektoen et al. (1995)
	Marine sediment (5–7 cm depth)	>300	
	Marine sediment	9–419	Bjorklund et al. (1990)
	Marine sediment	16	Coyne et al. (1994)
	Marine sediment	125	Hansen et al. (1993)
	Marine sediment	30–64	Samuelsen (1989)
	Marine sediment	70	Jacobsen and Berglind (1988)
	Marine sediment	87–144	Samuelsen et al. (1992)
	Marine sediment	41–83	Pouliquen et al. (1992)
	Marine sediment	No degradation after 180 d	Samuelsen et al. (1994)
	Sediment (aerobic)	<47 d	Lai et al. (1995)
	Sediment (anaerobic)	No degradation after 70 d	
	Photodegradation in water	96% degraded after 9 d	Lunestad et al. (1995)

Table 7. (Continued).

Compound	Test matrix/system	Half-life (d)	Reference
	Freshwater sediment (aerobic)	Rate 1.5–3.0 mg L^{-1} d^{-1}	Bohm (1996)
	Marine sediment (aerobic)	Rate 1.3–2.7 mg L^{-1} d^{-1}	
Oxolinic acid	Marine sediment (0–1 cm depth)	151	Hektoen et al. (1995)
	Marine sediment (5–7 cm depth)	>300	
	Marine sediment	165	Hansen et al. (1993)
	Marine sediment	48	Samuelsen (1992a)
	Marine sediment	No degradation after 180 d	Samuelsen et al. (1994)
	Photodegradation in water	88% degraded after 9 d	Lunestad et al. (1995)
Penicillin	OECD 301 D	36% degraded after 40 d	Al-Ahmad et al. (1999)
	Mixture of soil and chicken manure	<3 h	Gavalchin and Katz (1994)
Propetamphos	Hydrolysis pH 3	11	Reported in Lewis (1998)
	Hydrolysis pH 6	1 yr	
	Hydrolysis pH 9	41	
	Photolysis in aqueous solution	5	
Sarafloxicin	Marine sediment (0–1 cm depth)	151	Hektoen et al. (1995)
	Marine sediment (5–7 cm depth)	>300	
	Marine sediment	0.06% degraded after 83 d	Marengo et al. (1997)
	Loam soil	87–92% degraded after 80 d	
	Silt loam soil	82–89% degraded after 80 d	
	Sandy loam soil	69–82% degraded after 80 d	Velagaleti et al. (1993)
	Loam soil	0.66% degraded after 65 d	
	Silty clay loam soil	0.43% degraded after 65 d	
	Sandy clay loam soil	0.40% degraded after 65 d	
	Photodegradation in water	< 1 hr	Davis et al. (1993)
Sulfachloropyrazine	Laying hen feces	71% degraded after 3 mon	van Dijk and Keukens (2000)
	Broiler feces	65% degraded after 8 d	

Table 7. (*Continued*).

Compound	Test matrix/system	Half-life (d)	Reference
Sulfadiazine	Marine sediment (0–1 cm depth)	50	Hektoen et al. (1995)
	Marine sediment (5–7 cm depth)	100	
	Marine sediment	No degradation after 180 d	Samuelsen et al. (1994)
	Photodegradation in water	26% degraded after 21 d	Lunestad et al. (1995)
Sulfamethoxazole	OECD 301 D	No degradation after 40 d	Al-Ahmad et al. (1999)
Sulfadimethoxine	Marine sediment	20% degraded after 180 d	Samuelsen et al. (1994)
	Photodegradation in water	18% degraded after 21 d	Lunestad et al. (1995)
Tetracycline	Photodegradation in water	3 hr	Oka et al. (1989)
Trimethoprim	Marine sediment (0–1 cm depth)	75	Hektoen et al. (1995)
	Marine sediment (5–7 cm depth)	100	
	Marine sediment	<60 d	Samuelsen et al. (1994)
	Photodegradation in water	No degradation over 42 d	Lunestad et al. (1995)
Tylosin	Mixture of soil and chicken feces	<5 d	Galvachin and Katz (1994)
	Clay soil	3.3–8.1	
	Sandy soil	4.1–4.2	Ingerslev and Halling-Sørensen (2001)
	Pig slurry	<2	Loke et al. (2000)
Virginamycin	Sandy silt soil	40% mineralized after 64 d	Weerasinghe and Towner (1997)
	Silty sand soil	30% mineralized after 64 d	
	Silty sand soil	25% mineralized after 64 d	
	Silty clay loam soil	21% mineralized after 64 d	
	Clay loam soil	18% mineralized after 64 d	
	Silty clay loam soil	12% mineralized after 12 d	

OM: organic matter.

VII. Environmental Hazard (Continued)
A. Aquatic Toxicity

A large body of data is available on the aquatic toxicity of veterinary medicines covering a range of species and endpoints. Generally the available data cover three main therapeutic groups: sheep dip chemicals, antibacterial agents, and endectocides. Data are also available on the effects of the hormone treatment ethinyl estradiol.

Antiparasitics The effects of currently and previously approved sheep dip chemicals (including chlorfenvinphos, coumaphos, cypermethrin, deltamethrin, diazinon, fenchlorphos, flumethrin, and propetamphos) on aquatic organisms have been extensively investigated. Acute toxicity values for the compounds to insects, crustaceans, and fish are generally in the low ng L^{-1} to the low µg L^{-1} range, indicating a very high acute toxicity.

Few data are available on the toxicity of other ectoparasiticides to aquatic species. However, the aquatic toxicities of hydrogen peroxide and ivermectin have been extensively investigated because of their potential use in aquaculture. Avermectins are particularly toxic to crustaceans with effect levels (LC_{50}) to mysid shrimps ranging from 0.0026 µg L^{-1} (ivermectin) to 22 µg L^{-1} (abamectin). In acute toxicity tests, hydrogen peroxide is shown to range from nonhazardous to toxic to various species of fish and aquatic invertebrates (USEPA 2001). Chronic and acute toxicity data for phosmet, an ectoparasiticide used for the treatment of mange and louse infestations in pigs, show the compound to be very hazardous to daphnids (EC_{50}; 0.0056 mg L^{-1}) and fish (LC_{50}; 0.07 mg L^{-1}) (Lewis and Bardon 1998). Small numbers of data were available on the aquatic toxicity of the endoparasitic wormers, abamectin, triclabendazole, fenbendazole, and levamisole, the coccidiostat dimetridazole, and the endectocide doramectin.

Antibacterial Compounds A number of studies have investigated the effects of antibacterial veterinary medicines (Holten Lützhøft et al. 1999; Lanzkey and Halling-Sørensen 1997; Wollenberger et al. 2000). The toxic effect data for antibacterial agents on most aquatic species are generally in the mg L^{-1} range (Lanzky and Halling-Sørensen 1997; Migliore et al. 1997b). The exceptions are algae and cyanobacteria, with certain species (e.g., *Microcystis aruginosa*) being particularly sensitive, with reported EC_{50} values ranging from 0.0037 (amoxicillin) to 112 mg L^{-1} (trimethoprim), and the marine bacterium *Vibrio fischeri*, for which the toxicity ranged from 0.014 (ofloxacin) to 8.21 (streptomycin) mg L^{-1} (Backhaus and Grimme 1999).

A limited number of data were also available on the chronic toxicity of antibacterial compounds to daphnids (Wollenberger et al. 2000). Ratios of acute EC_{50}s or LOECs to chronic EC_{50}s or NOECs range from 2.2 to 16 (Table 10).

Table 8. Aquatic ecotoxicity data for a range of veterinary medicines.

Compound	Test organism	Toxic effect	Concentration (mg L^{-1})	Reference
Abamectin	Crassostrea virginica	96-hr LC$_{50}$	430	Reported in Davies et al. (1997)
	Daphnia magna	48-hr LC$_{50}$	0.34	
	Panaeus duorarum	96-hr LC$_{50}$	1.6	
	Mysidopsis bahia	96-hr LC$_{50}$	0.022	
	Callenectes sapidus	96-hr LC$_{50}$	153	
Aminosidine	D. magna	24-hr LC$_{50}$	1055	Reported in Holten Lützhøft et al. (1999)
	D. magna	48-hr EC$_{50}$	503	
	D. magna	Phototactic behavior increased	10	
	Artemia (nauplii)	48-hr EC$_{50}$	2220	Migliore et al. (1997a)
	Artemia (nauplii)	72-hr EC$_{50}$	846.5	
Amoxillin	Microcystis aeruginosa	EC$_{50}$	0.0037	Holten Lützhøft et al. (1999)
	Selenastrum capricornutum	NOEC	250	
	Rhodomonas salina	EC$_{50}$	3108	
Amitraz	Acanthocyclops vernalis	96-hr EC$_{50}$	58.5	USEPA (2001)
	Ankistiodesmus falcatus	10-d EC$_{50}$ (histology)	1.9	
	Anabaena flosaquae	5-d EC$_{50}$	3.4	
	Asellus brevicaudis	48-hr LC$_{50}$	>100	
	Chlorella vulgaris	96-hr NOEC	100	
	C. virginica	14-d LC$_{50}$	255.44	
	Cypidopsis vidua	48-hr LC$_{50}$	32	
	Cyprinodon variegatus	96-hr LC$_{50}$	>1000	
	D. magna	48-hr EC$_{50}$	18–30	
	Gammarus fasciatus	48-hr LC$_{50}$	100	
	Lepomis macrochirus	96-hr LC$_{50}$	10–1000	
	Notropis atherinoides	96-hr LC$_{50}$	420	
	Oncorhynchus mykiss	96-hr LC$_{50}$	243	
	Pimephales promelas	96-hr LC$_{50}$	100	
	Poecilia reticulata	96-hr LC$_{50}$	410	
	S. capricornutum	5-d EC$_{50}$	2.3	
	S. capricornutum	96-hr NOEC	1	

Table 8. (Continued).

Compound	Test organism	Toxic effect	Concentration (mg L^{-1})	Reference
Amitraz	Unspecified algae	72-hr EC$_{50}$	12	Lewis and Bardon (1998)
	D. magna	48-hr EC$_{50}$	0.035	
	Unspecified fish	96-hr LC$_{50}$	0.74	
Ampicillin	Vibrio fischeri	24-hr EC$_{50}$	163	Backhaus and Grimme (1999)
Apramycin	Rainbow trout	96-hr LC$_{50}$	>300	Elanco (2000)
	Bluegill sunfish	96-hr LC$_{50}$	>300	
	D. magna	48-hr EC$_{50}$	101.6	
Azamethiphos	Homarus americanus	48-hr LC$_{50}$	0.00103–0.00357	USEPA (2001)
	Lepeophtheirus salmonis	40 min (behavior)	0.1	
	Lepeophtheirus salmonis	40 min (increased mortality)	0.1	
Bacitracin	Artemia salina (nauplii)	24-hr EC$_{50}$	34.1	Migliore et al. (1997a)
	A. salina (nauplii)	48-hr EC$_{50}$	21.8	
	A. salina	LC$_{100}$	6.3	
	A. salina (cysts)	Hatching	25	
	D. magna	24-hr LC$_{50}$	126.4	Reported in Webb (2001)
	D. magna	48-hr LC$_{50}$	30.5	Reported in Holten Lützhøft et al. (1999)
	D. magna	Phototactic behavior decreased	10	
Benzyl alcohol	D. magna	24-hr EC$_{50}$ (behavior)	55–400	USEPA (2001)
	Haematococcus pluvialis	4-hr EC$_{50}$ (histology)	2600	
	Leuciscus idus melanotus	48-hr LC$_{50}$	646	
	Aedes aegypti	24-hr EC$_{50}$	105–129	
	Aedes scutellaris	24-hr EC$_{50}$	110–126	
	L. machrochirus	96-hr LC$_{50}$	10	
	Mendia beryllina	96-hr LC$_{50}$	15	
	P. promelas	96-hr LC$_{50}$	460	
	Tetrahymena pyriformis	48-hr EC$_{50}$	853	
	Petromyzon marinus	24 EC$_{50}$	5	

Table 8. (Continued).

Compound	Test organism	Toxic effect	Concentration (mg L^{-1})	Reference
Benzyl penicillin	*M. aeruginosa*	7-d EC$_{50}$	0.006	Halling-Sørensen (2000)
	S. capricornutum	72-hr NOEC	100	
Chloramphenicol	*A. salina*	24-hr LC$_{50}$	2042	Reported in Webb (2001)
	Brachionus calyciflorus	24-hr LC$_{50}$	2074	
	D. magna	24-hr EC$_{50}$	543	
	D. magna	24-hr EC$_{50}$	1086	
	Streptocephalus proboscideus	24-hr LC$_{50}$	305	
	Vibrio fischeri	24-hr EC$_{50}$	0.0643	Backhaus and Grimme (1999)
Chlorfenvinphos	*Scenedesmus vacuolatus*	Inhibition of reproduction	4.07	Meyer et al. (2001)
	S. quadricauda	24-hr EC$_{100}$ (photosynthesis)	100	Reported in Lewis et al. (1993)
	Lemna minor	9–10 d interruption of chlorophyll	0.6	
	D. magna	24-hr LC$_{50}$	0.00028	
	D. magna	48-hr LC$_{50}$	0.0001	
	G. fasciatus	24-hr LC$_{50}$	0.027	
	G. fasciatus	96-hr LC$_{50}$	0.0096	
	Pteronarcys californica	24-hr LC$_{50}$	0.0058	
	Pteronarcyc californica	96-hr LC$_{50}$	0.0007	
	Culex pipiens	72-hr LC$_{100}$	0.002	
	Carp	24-hr LC$_{50}$	0.055	
	Carp	48-hr LC$_{50}$	0.045–0.27	
	Lebistes reticulatus	24-hr LC$_{50}$	2.1	
	L. reticulatus	48-hr LC$_{50}$	1.78	
	L. reticulatus	96-hr LC$_{50}$	1.5	
	L. reticulatus (juvenile)	24-hr LC$_{50}$	1.4–2.7	
	L. reticulatus (juvenile)	24-hr symptoms of poisoning	0.5	

Table 8. (Continued).

Compound	Test organism	Toxic effect	Concentration (mg L^{-1})	Reference
	L. macrochirus	24-hr LC$_{50}$	0.0028–0.05	
	L. macrochirus	96-hr LC$_{50}$	0.023	
	P. reticulata	24-hr LC$_{50}$	2.03	
	P. reticulata	48-hr LC$_{50}$	0.53	
	P. reticulata	72–96 hr LC$_{50}$	1.56	
	Rasbora heteromorpha	24-hr LC$_{50}$	0.36	
	R. heteromorpha	48-hr LC$_{50}$	0.27	
	R. heteromorpha	96-hr LC$_{50}$	0.32	
	O. mykiss	24-hr LC$_{50}$	1.65	
	O. mykiss	96-hr LC$_{50}$	0.51	
	Carrassius auratus	48-hr LC$_{50}$	0.34	
	Scenedesmus subspicatus	72-hr EC$_{50}$ (inhibition of growth)	1.94	Reported in Lewis (1998)
	S. subspicatus	96-hr EC$_{50}$ (inhibition of growth)	1.36	
	S. subspicatus	96-hr NOEC (inhibition of growth)	0.246	Reported in Lewis (1998)
	S. subspicatus	96-hr LOEC (inhibition of growth)	0.788	
	S. capricornutum	96-hr EC$_{50}$ (inhibition of growth)	1.6	
	D. magna	24-hr EC$_{50}$ (immobilisation)	0.0012	
	D. magna	48-hr EC$_{50}$ (immobilisation)	0.00025	
	D. magna	24-hr EC$_{50}$	0.0018	
	D. magna	48-hr EC$_{50}$	0.00046	
	Ceriodaphnia dubia	48-hr LC$_{50}$	0.0004	
Chlortetracycline	M. aeruginosa	7-d EC$_{50}$	0.05	Halling-Sørensen (2000)
	S. capricornutum	72-hr EC$_{50}$	3.1	Halling-Sørensen (2000)

Table 8. (Continued).

Compound	Test organism	Toxic effect	Concentration (mg L^{-1})	Reference
Cinoxacin	V. fischeri	24-hr EC$_{50}$	0.117	Backhaus et al. (2000)
	S. vacuolatus	EC$_{50}$	>26	Backhaus et al. (2001)
	Anguilla japonica	LC$_{50}$	73	USEPA (2001)
Ciprofloxacin	M. aeruginosa	EC$_{50}$	0.005	Holten Lützhøft and Halling-Sørensen (unpublished)
	Pseudomonas putida	EC$_{50}$ (growth inhibition)	0.08	Kümmerer et al. (2000)
Coumaphos	Gammarus lacustris (scud)	24-hr LC$_{50}$	0.00032	Reported in Lewis et al. (1993)
	G. lacustris (scud)	48-hr LC$_{50}$	0.00014	
	G. lacustris (scud)	96-hr LC$_{50}$	0.000074	
	G. lacustris (shrimp)	48-hr LC$_{50}$	0.002	
	Simocephalus serrulatus (1st instar)	48-hr LC$_{50}$	0.0001	
	S. serrulatus	48-hr LC$_{50}$	0.001	
	Hexagenia spp. (naiad)	24-hr LC$_{50}$	0.43	
	Hydropsyche spp. (larvae)	24-hr LC$_{50}$	0.005	
	O. mykiss (juvenile)	24-hr LC$_{50}$	2.6–3.0	
	O. mykiss (juvenile)	48-hr LC$_{50}$	0.55–1.8	
	O. mykiss	72-hr LC$_{50}$	1.5	
	O. mykiss	96-hr LC$_{50}$	0.89–1.5	
	Salmo trutta	24-hr LC$_{50}$	0.92	
	S. trutta	48-hr LC$_{50}$	0.73	
	Salvelinus fontinalis	24-hr LC$_{50}$	1.06	
	S. fontinalis	48-hr LC$_{50}$	0.8	
	Salvelinus namaycush (juvenile)	24-hr LC$_{50}$	6.8	
	S. namaycusk (juvenile)	48-hr LC$_{50}$	4	Reported in Lewis et al. (1993)
	S. namaycush	24-hr LC$_{50}$	0.99	
	S. namaycush	96-hr LC$_{50}$	0.59	
	Onchorhyncus clarkii	24-hr LC$_{50}$	1.09	

Table 8. (*Continued*).

Compound	Test organism	Toxic effect	Concentration (mg L^{-1})	Reference
	O. clarkii	96-hr LC$_{50}$	0.86	
	Onchorhyncus kisutch	24-hr LC$_{50}$	22	
	O. kisutch	48-hr LC$_{50}$	20	
	O. kisutch	72-hr LC$_{50}$	18	
	O. kisutch	96-hr LC$_{50}$	15	
	Ictalurus punctatus (juvenile)	24-hr LC$_{50}$	6.8	
	I. punctatus	24-hr LC$_{50}$	5.2	
	I. punctatus	96-hr LC$_{50}$	0.84	
	L. macrochirus (juvenile)	24-hr LC$_{50}$	10.5	
	L. macrochirus (juvenile)	48-hr LC$_{50}$	8	
	L. macrochirus	24-hr LC$_{50}$	1.1–1.4	
	L. macrochirus	96-hr LC$_{50}$	0.18–0.34	
	P. promelas	96-hr LC$_{50}$	>18	
	P. promelas	48-hr LC$_{50}$	>1	
	Micropterus salmoides	24-hr LC$_{50}$	1.5	
	M. salmoides	36-hr LC$_{50}$	0.5	
	M. salmoides	96-hr LC$_{50}$	1.1	
	L. reticulatus	96-hr LC$_{50}$	>0.56	
	R. heteromorpha	24-hr LC$_{50}$	0.082	
	R. heteromorpha	48-hr LC$_{50}$	0.046	
	Carassius auratus	96-hr LC$_{50}$	>18	
	Stizostedion vitreum	24-hr LC$_{50}$	1.35	
	S. vitreum	96-hr LC$_{50}$	0.78	
	A. salina [larvae (24 hr)]	24-hr LC$_{50}$	21.23	
	A. salina [larvae (48 hr)]	24-hr LC$_{50}$	5.51	
	A. salina [larvae (72 hr)]	24-hr LC$_{50}$	5.22	Reported in Lewis (1998)
Cypermethrin	*D. magna*	48-hr LC$_{50}$	0.00015	Tomlin (1997)
Cyromazine	*D. magna*	48-hr EC$_{50}$	97.8	USEPA (2001)
	Deleatidium spp.	48-hr LC$_{50}$	>300	

Table 8. (Continued).

Compound	Test organism	Toxic effect	Concentration (mg L^{-1})	Reference
	Gambusia affinis	72-hr LC$_{50}$	0.037	Tomlin (1997)
	I. punctatus	95-hr LC$_{50}$	91.6	USEPA (2001)
	L. macrochirus	96-hr LC$_{50}$	89.7	
	O. mykiss	96-hr LC$_{50}$	87.9	
	Dugesia dorotocephala	72-hr LC$_{50}$	>10	
	Dugesia tigrina	72-hr LC$_{50}$	>10	
	D. tigrina	72-hr EC$_{50}$ (reproduction)	>10	
Deltamethrin	D. magna	48-hr LC$_{50}$	0.0035	
	Aedes aegypli	96-hr LC$_{50}$	0.00015	
	Alburnus alburnus	7-d LC$_{50}$	8.8	
	Americamysis bahia	96-hr LC$_{50}$	0.0017–0.0037	
	Anadonta anatina	7-d LC$_{50}$	8600	
	Anadonta cygnea	7-d LC$_{50}$	6300	
	Baetis parvus	1-hr LC$_{50}$	0.0004	
	Bufo arenarium	96-hr LC$_{50}$	0.0045	
	Chironomus decorus	24-hr LC$_{50}$	0.00027–0.0011	
	Chironomus salinarius	24-hr LC$_{50}$	0.00071	
	Chironomus utahensis	24-hr LC$_{50}$	0.00029	
	C. virginica	96-hr EC$_{50}$	17.9–110	
	C. virginica	96-hr EC$_{50}$	0.018–0.11	
	Cricotopus	24-hr LC$_{50}$	0.00011–0.00015	
	Ctenopharyngodon idella	96-hr LC$_{50}$	0.091	
	C. pipiens	24-hr LC$_{50}$	0.00002	
	Cyprinodon macularius	48-hr LC$_{50}$	0.0006	
	Cyprinodon variegatus	96-hr LC$_{50}$	0.00036–0.00058	
	Cyprinus carpio	96-hr LC$_{50}$	0.078	
	Homarus americanus	96-hr LC$_{50}$	0.0000014	
	L. macrochirus	96-hr LC$_{50}$	0.00036–0.0015	
	Lymnaea acuminata	96-hr LC$_{50}$	0.44–0.45	

Table 8. (Continued).

Compound	Test organism	Toxic effect	Concentration (mg L^{-1})	Reference
	O. mykiss	96-hr LC$_{50}$	0.00025–0.0023	
	S. salar	96-hr LC$_{50}$	0.00197	
	Procladius	24-hr LC$_{50}$	0.000067	
	Similium virgatum	1-hr LC$_{50}$	0.0009	
	Tanypus nubifer	24-hr LC$_{50}$	0.00011	
	Tilapia mossambica	48-hr LC$_{50}$	0.0008	
	Tilipia nilotica	96-hr LC$_{50}$	0.000145	
Diazinon	Gillia altilis	96-hr LC$_{50}$	11	Reported in Lewis et al. (1993)
	Daphnia sp.	48-hr LC$_{50}$	0.0009–0.0018	
	D. magna	48-hr LOEC	0.0015	
	D. pulex	48-hr EC$_{50}$	0.0008	Burkepile et al. (2000)
	G. fasciatus	24-hr LC$_{50}$	0.008	Reported in Lewis et al. (1993)
	G. fasciatus	96-hr LC$_{50}$	0.0002	
	G. lacustris	24-hr LC$_{50}$	0.8	
	G. lacustris	48-hr LC$_{50}$	0.5	
	G. lacustris	96-hr LC$_{50}$	0.2	
	S. serrulatus	48-hr EC$_{50}$	0.0014–0.0018	
	P. californica	24-hr LC$_{50}$	0.155	
	P. californica	48-hr LC$_{50}$	0.06	
	P. californica	96-hr LC$_{50}$	0.025	
	O. mykiss (juvenile)	24-hr LC$_{50}$	0.38	
	O. mykiss (juvenile)	48-hr LC$_{50}$	0.17	
	O. mykiss (juvenile)	96-hr LC$_{50}$	0.09	
	O. mykiss	96-hr LC$_{50}$	0.02–3.2	
	S. namaycush	24-hr LC$_{50}$	2.19	
	S. namaycush	96-hr LC$_{50}$	0.6	
	O. clarkii	24-hr LC$_{50}$	2.58–3.59	
	O. clarkii	96-hr LC$_{50}$	1.7–2.76	
	S. fontinalis	96-hr LC$_{50}$	0.77	

Table 8. (Continued).

Compound	Test organism	Toxic effect	Concentration (mg L^{-1})	Reference
	L. macrochirus (juvenile)	24-hr LC$_{50}$	0.052–0.36	
	L. macrochirus (juvenile)	48-hr LC$_{50}$	0.03	
	L. macrochirus (juvenile)	96-hr LC$_{50}$	0.02–0.17	
	L. macrochirus	48-hr LC$_{50}$	0.08	
	L. macrochirus	96-hr LC$_{50}$	0.09–16	
	P. promelas (juvenile)	96-hr LC$_{50}$	7.8	
	Jordinella floridae	96-hr LC$_{50}$	1.6	
	R. heteromorpha	24-hr LC$_{50}$	1.45	
	Cyprinus carpio	96-hr LC$_{50}$	7.6–23.4	
	Channa punctatus	96-hr LC$_{50}$	3.1	
	Saccobranchus fossilis	24-hr LC$_{50}$	5.14	
	S. fossilis	96-hr LC$_{50}$	4.57	
	Brachydanio rerio	24-hr LC$_{50}$	2.3	
	B. rerio	48-hr LC$_{50}$	2.24	
	B. rerio	72-hr LC$_{50}$	2.19	
	B. rerio	96-hr LC$_{50}$	2.12	
	Ceriodaphnia dubia	48-hr LOEC	0.0008	Burkepile et al. (2000)
	Hyallela azteca	48-hr LOEC	0.011	
	Chironomus tentans	48-hr LOEC	0.0375	
	P. promelas	48-hr LOEC	12.5	
	Oryzias latipes	48-hr LC$_{50}$	4.4	Reported in Larkin and Tjeerdema (2000)
	Anguilla anguilla	96-hr LC$_{50}$	0.8	
	Cyprinodon variegatus	96-hr LC$_{50}$	1.4	
	Ceriodaphnia dubia	48-hr LC$_{50}$	0.000026–0.00058	
	D. tigrina	96-hr LC$_{50}$	0.63 + 0.2	
	Brachionus calyciflorus	24-hr LC$_{50}$	29.22	
	Chironomus tepperi (4th instar larvae)	LC$_{50}$	0.0355	
	M. bahia (juvenile)	96-hr LC$_{50}$	0.00858	
	Penaeus duorarum (post larvae)	96-hr LC$_{50}$	0.021	

Table 8. (*Continued*).

Compound	Test organism	Toxic effect	Concentration (mg L^{-1})	Reference
Dichlorvos	*Daphnia* spp.	48-hr LC$_{50}$	0.00019	Tomlin (1997)
Doramectin	*D. magna*	48-hr NOEC	0.025	Taylor (1999)
	L. macrochirus	96-hr LC$_{50}$	11	
	Salmo gairdneri	96-hr LC$_{50}$	5.1	
Dimetridazole	*Gyrodactylus salaris*	25% mortality over 1 hr	200	USEPA (2001)
	O. mykiss	Behavioral effect	200	
Enoxacin	*V. fischeri*	24-hr EC$_{50}$	0.049	Backhaus et al. (2000)
	S. vacuolatus	EC$_{50}$	19.7	
Enrofloxacin	Rainbow trout	96-hr LC$_{50}$	>10	Bayer (1997)
	L. macrochirus	96-hr LC$_{50}$	>10	
	D. magna	24-hr LC$_{50}$	>10	
Eprinomectin	*D. magna*	48-hr EC$_{50}$	0.00045	Merial (1998)
Erythromycin	*D. magna*	24-hr LC$_{50}$	388	Reported in Webb (2001)
	D. magna	48-hr LC$_{50}$	211	
	D. magna	48-hr EC$_{50}$	30.5	Reported in Holten Lützhøft et al. (1999)
	Artemia (cysts)	120-hr LC$_{100}$	<10	Migliore et al. (1997a)
	Artemia (cysts)	48-hr NOEC	<10	Migliore et al. (1997a)
Erythromycin	*D. magna*	24-hr EC$_{50}$	5.7	Reported in Webb (2001)
	D. magna	48-hr EC$_{50}$	6.4	
	O. mykiss	96-hr EC$_{50}$	1.6	
	P. putida	Microbial growth inhibition	>20	Schweinfurth et al. (1996)
	Azobacter beijerincki	Microbial growth inhibition	>20	
	Aspergillus niger	Microbial growth inhibition	>20	
	Chaetomium globosum	Microbial growth inhibition	>20	
	Nostoc ellipsporum	Microbial growth inhibition	>20	
Ethinyl estradiol	Unspecified algae	EC$_{50}$	0.84	Kopf (1995)

Table 8. (Continued).

Compound	Test organism	Toxic effect	Concentration (mg L^{-1})	Reference
	Daphnia spp.	EC$_{50}$ (reproduction)	0.105	Schweinfurth et al. (1996)
	Daphnia spp.	EC$_{50}$ (acute test)	5.7	
	Daphnia spp.	48-hr EC$_{50}$	6.4	
	Daphnia spp.	NOEC (immobilization)	3	
	Daphnia spp.	21-d NOEC (number of offspring)	>0.387	
	Daphnia spp.	21-d NOEC (immobilization)	>0.387	
	O. mykiss	96-hr EC$_{50}$	1.61	
	P. promelas (larvae)	28-d LOEC (changes to kidney/liver)	0.00001	
	P. promelas (larvae)	28-d LOEC (decreased growth)	≥0.0001	
	P. promelas (larvae)	28-d LOEC (mortality)	≥0.001	
	P. promelas (juvenile)	28-d LOEC (changes to kidney/liver)	0.00001	
	P. promelas (juvenile)	28-d LOEC (mortality)	≥0.001	
	P. promelas (adult)	28-d LOEC (inhibited egg production)	≥0.00001	
	P. promelas (adult)	28-d LOEC (mortality)	≥0.001	
	P. promelas	9-mon reproduction NOEC	0.000001	
	Rutilus rutilus	10-d plasma vitellogenin NOEC (9 °C)	0.000001	Reported in Webb (2001)
	O. mykiss	10-d plasma vitellogenin NOEC (9 °C)	0.000001	
	O. mykiss	28-wk plasma vitellogenin LOEC	0.0000003	
	O. mykiss	10-d plasma vitellogenin LOEC (16.5 °C)	0.0000001	
	Lymnaea stagnalis	50–60 d LOEC (growth)	0.00000125	Reported in Webb (2001)

Table 8. (Continued).

Compound	Test organism	Toxic effect	Concentration (mg L^{-1})	Reference
Fenbendazole	Pseudodactylogyrus	Population decrease over 24 hr	1–10	USEPA (2001)
	Anguilla anguilla	24 hr (physiology and growth effect)	1–10	Lewis and Bardon (1998)
	D. magna	48-hr EC$_{50}$	1000	
	Unspecified fish	96-hr LC$_{50}$	500	
Fenchlorphos	Unspecified shrimp	48-hr LC$_{50}$	0.005	Reported in Lewis et al. (1993)
	I. punctatus (juvenile)	24-hr LC$_{50}$	1.76	
	I. punctatus (juvenile)	48-hr LC$_{50}$	1.26	
	L. macrochirus (juvenile)	24-hr LC$_{50}$	2.5	
	L. macrochirus (juvenile)	48-hr LC$_{50}$	1	
	O. mykiss (juvenile)	24-hr LC$_{50}$	1.17	
	O. mykiss (juvenile)	48-hr LC$_{50}$	0.74	
	S. trutta (juvenile)	24-hr LC$_{50}$	0.53	
	S. trutta (juvenile)	48-hr LC$_{50}$	0.39	
	S. fontinalis (juvenile)	24-hr LC$_{50}$	0.59	
	S. fontinalis (juvenile)	48-hr LC$_{50}$	0.39	
	S. namaycush (juvenile)	24-hr LC$_{50}$	0.73	
	S. namaycush (juvenile)	48-hr LC$_{50}$	0.62	
Flumequine	M. aeruginosa	EC$_{50}$	0.159	Holten Lützhøft et al. (1999)
	S. capricornutum	EC$_{50}$	5.0	
	R. salina	EC$_{50}$	18	
	Lythrum salicaria	Growth	<100	Migliore et al. (2000)
	Vibrio fischeri	24-hr EC$_{50}$	0.019	Backhaus et al. (2000)
	A. salina (nauplii)	24-hr EC$_{50}$	477	Migliore et al. (1997a)
	A. salina (nauplii)	48-hr EC$_{50}$	308	
	A. salina (nauplii)	72-hr EC$_{50}$	96.4	
	A. salina (nauplii)	22% mortality	6.31	
	S. vacuolatus	EC$_{50}$	3.7	Backhaus et al. (2001)

Table 8. (Continued).

Compound	Test organism	Toxic effect	Concentration (mg L^{-1})	Reference
Furazolidone	A. salina	LC$_{50}$	250	Reported in Holten Lützhøft et al. (1999)
	D. magna	LC$_{50}$	60	
	C. pipiens (larvae)	LC$_{50}$	40	
Griseofulvin	D. magna	24-hr EC$_{50}$ (physiology)	>1000	USEPA (2001)
	D. magna	48-hr EC$_{50}$ (physiology)	>1000	
	D. magna	72-hr EC$_{50}$ (physiology)	>1000	
	Mercenaria mercenaria	48-hr EC$_{50}$ (developmental)	<0.25	
	M. mercenaria	14-d LC$_{50}$	<1	
Halothane	Lymnaea stagnalis	Behavioral effects	0.5%	USEPA (2001)
Hydrogen peroxide	D. magna	24-hr EC$_{50}$	2.3	USEPA (2001)
	A. salina	24-hr LC$_{50}$	918	
	L. macrochirus	96-hr LC$_{50}$	26.7	
	O. mykiss	96-hr LC$_{50}$	22	
	Siganus fuscescens	24-hr LC$_{50}$	224	
	Trachurus japonicus	24-hr LC$_{50}$	89	
	Tridentiger trigonocephalus	24-hr LC$_{50}$	155	
Ivermectin	Asterias rubens (sediment test)	10-d LC$_{50}$	23.6 mg kg^{-1}	Davies et al. (1998)
	C. volutator (sediment test)	10-d LC$_{50}$	0.18 mg kg^{-1}	
	A. marina	10-d LC$_{50}$	0.018 mg kg^{-1}	Thain et al. (1997)
	A. marina	Effects on feeding	<0.005 mg kg^{-1}	
	A. marina	Adverse effect on burrowing	>0.008 mg kg^{-1}	
	S. gardneiri	96-hr LC$_{50}$	0.003	Halley et al. (1989)
	L. macrochirus	96-hr LC$_{50}$	0.0048	
	Crangon septemspinosa	96-hr LC$_{50}$	0.021	Reported in Davies et al. (1997)
	Neomysis integer	96-hr LC$_{50}$	0.07	Davies et al. (1997)
	N. integer	48-hr LC$_{50}$	0.000026	Grant and Briggs (1998)

Table 8. (Continued).

Compound	Test organism	Toxic effect	Concentration (mg L^{-1})	Reference
	Gammarus spp.	96-hr LC$_{50}$	0.000033	
	Palaemonectes varians	96-hr LC$_{50}$	0.054	
	A. salina	24-hr LC$_{50}$	>0.03	
	Sphaeroma rugicauda	96-hr LC$_{50}$	0.348	
	Carcinus maenas	96-hr LC$_{50}$	0.957	Davies and Rodger (2000)
	Crassostrea gigas (larvae)	96-hr LC$_{50}$	80–100	
	C. gigas (spat)	96-hr LC$_{50}$	460	
	Mytilus edulis	96-hr LC$_{50}$	400	
	Tapes semidecassata (larvae)	96-hr LC$_{50}$	0.38	
	Tapes semidecassata (spat)	96-hr LC$_{50}$	0.60	
	Pecten maximus	96-hr LC$_{50}$	0.30	Davies and Rodger (2000)
	Monodonta lineata	96-hr LC$_{50}$	0.78	
	Nucella lapillus	96-hr LC$_{50}$	0.39	
	Littorina littorea	96-hr LC$_{50}$	0.58	
	Hydrobia ulvae	96-hr LC$_{50}$	>10	
	Potamopyrgus jenkinsii	96-hr LC$_{50}$	<9	Grant and Briggs (1998)
	Nereis diversicolor	96-hr LC$_{50}$	0.0075	
	A. marina	10-d LC$_{50}$ (sediment)	0.023 mg kg^{-1}	
	Biomphalaria glabrata	24-hr LC$_{50}$	0.03	Matha and Weiser (1988)
	D. magna	48-hr LC$_{50}$	0.000025	Halley et al. (1989)
	Chlorella pyrenoidosa	14-d LC$_{50}$	>10000	
Levamisole	A. anguilla	88% physiological effect over 25 hr	10	USEPA (2001)
Lincomycin	D. magna	48-hr EC$_{50}$	379.4	Reported in Holten Lützhøft et al. (1999)
	D. magna	Phototactic behavior decreased	5	
	Artemia spp.	72-hr EC$_{50}$	283	Migliore et al. (1997a)

Table 8. (Continued).

Compound	Test organism	Toxic effect	Concentration (mg L^{-1})	Reference
Lomefloxacin	Daphnia spp.	EC$_{50}$	130	Reported in Webb (2001)
	O. mykiss	LC$_{50}$	170	
	Unspecified green algae	EC$_{50}$	2.4	
	Unspecified green algae	NOEC	2	
	V. fischeri	24-hr EC$_{50}$	0.022	Backhaus et al. (2000)
	S. vacuolatus	EC$_{50}$	58	
Metronidazole	S. capricornutum	72-hr EC$_{50}$	39.1–40.4	Lanzky and Halling-Sørensen (1997)
	S. capricornutum	72-hr EC$_{10}$	19.9	Reported in Webb (2001)
	D. magna	48-hr LOEC	1000	Wollenberger et al. (2000)
	Chlorella spp.	72-hr EC$_{50}$	12.5–45.1	Lanzky and Halling-Sørensen (1997)
	Chlorella spp.	72-hr EC$_{10}$	2.03	Reported in Webb (2001)
	Acartia tonsa	72-hr EC$_{50}$	>100	Lanzky and Halling-Sørensen (1997)
	Brachydanio rerio	96-hr NOEC	>500	Kummerer et al. (2000)
	P. putida	EC$_{50}$ (growth inhibition)	>64	Reported in Webb (2001)
	O. mykiss	48-hr LC$_{50}$	>100	
	S. trutta	48-hr LC$_{50}$	>100	
	S. fontinalis	48-hr LC$_{50}$	>100	
	I. punctatus	48-hr LC$_{50}$	>100	
	L. machrochirus	48-hr LC$_{50}$	>100	
	S. namaycush	48-hr LC$_{50}$	>100	
Monensin	Rainbow trout	96-hr LC$_{50}$	9.0	Elanco (1998)
	Bluegill sunfish	96-hr LC$_{50}$	16.6	
	D. magna	48-hr EC$_{50}$	10.7	
Naladixic acid	V. fischeri	24-hr EC$_{50}$	0.200	Backhaus et al. (2000)
	S. vacuolatus	Inhibition of reproduction	21.9	Meyer et al. (2001)
	S. vacuolatus	EC$_{50}$	22.9	Backhaus et al. (2001)
Neomycin	A. japonica	LC$_{50}$	2829	USEPA (2001)
Norfloxacin	V. fischeri	24-hr EC$_{50}$	0.022	Backhaus et al. (2000)
	S. vacuolatus	EC$_{50}$	69.6	

Table 8. (Continued).

Compound	Test organism	Toxic effect	Concentration (mg L^{-1})	Reference
Ofloxacin	V. fischeri	24-hr EC$_{50}$	0.014	Backhaus et al. (2000)
	S. vacuolatus	EC$_{50}$	82.8	Backhaus et al. (2001)
	P. putida	EC$_{50}$ (growth inhibition)	0.01	Kummerer et al. (2000)
Olaquindox	D. magna	48-hr LOEC	1000	Wollenberger et al. (2000)
	M. aeruginosa	7-d EC$_{50}$	5.1	Halling-Sørensen (2000)
	S. capricornutum	72-hr EC$_{50}$	40	
Oxolinic acid	M. aeruginosa	EC$_{50}$	0.18	Holten Lützhøft et al. (1999)
	S. capricornutum	EC$_{50}$	16	
	R. salina	EC$_{50}$	10	
	D. magna	48-hr EC$_{50}$	4.6	Wollenberger et al. (2000)
	V. fischeri	24-hr EC$_{50}$	0.023	Backhaus et al. (2000)
	S. vacuolatus	EC$_{50}$	>26	Backhaus et al. (2001)
Oxytetracycline	M. aeruginosa	EC$_{50}$	0.207	Holten Lützhøft et al. (1999)
	S. capricornutum	EC$_{50}$	4.5	Holten Lützhøft et al. (1999)
	R. salina	EC$_{50}$	1.6	Holten Lützhøft et al. (1999)
	D. magna	48-hr LOEC	100	Wollenberger et al. (2000)
	D. magna	48-hr EC$_{50}$ intoxication	>102 ppm	USEPA (2001)
	L. macrochirus	96-hr LC$_{50}$	>100 ppm	USEPA (2001)
	Morone saxatilis (larvae)	24-hr LC$_{50}$	62.5	Reported in Webb (2001)
	M. saxatilis (larvae)	48-hr LC$_{50}$	62.5	
	M. saxatilis (larvae)	72-hr LC$_{50}$	62.5	
	M. saxatilis (larvae)	96-hr LC$_{50}$	62.5	
	M. saxatilis (fingerling)	24-hr LC$_{50}$	150	
	M. saxatilis (fingerling)	48-hr LC$_{50}$	125	
	M. saxatilis (fingerling)	72-hr LC$_{50}$	100	
	M. saxatilis (fingerling)	96-hr LC$_{50}$	75	
	O. mykiss	96-hr LC$_{50}$	>116 ppm	USEPA (2001)
	Panneus vannamei	24-hr EC$_{50}$ intoxication	0.16	
	P. vannamei	48-hr EC$_{50}$ intoxication	0.0611–0.2141	

Table 8. (*Continued*).

Compound	Test organism	Toxic effect	Concentration (mg L^{-1})	Reference
	P. vannamei	24-hr LC$_{50}$	0.16	Reported in Webb (2001)
	P. vannamei	48-hr LC$_{50}$	0.16–0.2384	
	P. vannamei	24-hr LOEC intoxication	0.1609	
	P. vannamei	48-hr LOEC intoxication	0.1089–0.3778	
	P. vannamei	24-hr NOEC intoxication	0.1609	
	P. vannamei	48-hr NOEC intoxication	0.0549–0.1609	
	S. namaycush	24-hr LC$_{50}$	<200	
Pipemidic acid	*V. fischeri*	24-hr EC$_{50}$	1.019	Backhaus et al. (2000)
	S. vacuolatus	EC$_{50}$	>151	
Pirimidic acid	*V. fischeri*	24-hr EC$_{50}$	0.121	Backhaus et al. (2000)
Phosmet	*D. magna*	48-hr EC$_{50}$	0.0056	Lewis and Bardon (1998)
	D. magna	Unspecified chronic test	0.0016	
	Unspecified fish	96-hr LC$_{50}$	0.07	
Procaine HCl	*L. macrochirus*	4-d (physiology)	77–101	USEPA (2001)
	Ptychocheilus spp.	24-hr (mortality)	10	
Propetamphos	*D. magna*	24-hr EC$_{50}$	0.0147	Reported in Lewis (1998)
	D. magna	48-hr EC$_{50}$	0.00878	
	Photobacterium phosporeum	30-min EC$_{50}$	21.4	
Salinomycin	Golden orfe	96-hr LC$_{50}$	32.2	Intervet (1999)
	Golden orfe	48-hr LC$_{50}$	27.5	
Sarafloxicin	*M. aeruginosa*	EC$_{50}$	0.015	Holten Lützhøft et al. (1999)
	R. salina	EC$_{50}$	24	
	S. capricornutum	EC$_{50}$	16	
Spiramycin	*M. aeruginosa*	7-d EC$_{50}$	0.005	Halling-Sørensen (2000)
	S. capricornutum	72-hr EC$_{50}$	2.3	

Table 8. (Continued).

Compound	Test organism	Toxic effect	Concentration (mg L^{-1})	Reference
Streptomycin	M. aeruginosa	MIC	0.3	Holten Lützhøft et al. (1999)
	S. capricornutum	MIC	2.1	Wollenberger et al. (2000)
	D. magna	48-hr EC$_{50}$	487	Halling-Sørensen et al. (2000)
	M. aeruginosa	7-d EC$_{50}$	0.007	Backhaus and Grimme (1999)
	S. capricornutum	72-hr EC$_{50}$	0.133	
	V. fischeri	24-hr EC$_{50}$	8.21	Meyer et al. (2000)
	S. vacuolatus	Inhibition of reproduction	17.4	
Sulfadiazine	M. aeruginosa	EC$_{50}$ population	0.135	Holten Lützhøft et al. (1999)
	S. capricornutum	EC$_{50}$	7.8	
	R. salina	EC$_{50}$	403	
	D. magna	48-hr EC$_{50}$	221	Wollenberger et al. (2000)
	D. magna	24-hr EC$_{50}$ physiology	112	USEPA (2001)
	D. magna	48-hr EC$_{50}$ physiology	88	
	D. magna	72-hr EC$_{50}$ physiology	57	
	Cirrhinus mrigala	Effect on growth	20 mg/100 g	
Sulfachloropyridazine	D. magna	48-hr EC$_{50}$	250	Novartis (1999)
	Zebrafish	96-hr LC$_{50}$	>1000	
Sulfadimethoxine	A. salina (nauplii)	24-hr LC$_{50}$	1866	Reported in Webb (2001)
	A. salina (nauplii)	48-hr LC$_{50}$	851	
	A. salina (nauplii)	72-hr LC$_{50}$	537	
	A. salina (nauplii)	96-hr LC$_{50}$	19.5	
Teflubenzuron	Fish (trout and carp)	96-hr LC$_{50}$	>500	Tomlin (1997)
Tetracycline	D. magna	48-hr NOEC	340	Wollenberger et al. (2000)
	M. aeruginosa	7-d EC$_{50}$	0.09	Halling-Sørensen (2000)
	S. capricornutum	72-hr EC$_{50}$	2.2	
	Nitzschia closterium	72-hr EC$_{50}$	16	Reported in Webb (2001)

Table 8. (*Continued*).

Compound	Test organism	Toxic effect	Concentration (mg L^{-1})	Reference
Tetracycline	*V. fischeri*	24-hr EC$_{50}$	0.0251	Backhaus and Grimme (1999)
	C. gigas	48-hr EC$_{50}$ developmental	81–89	USEPA (2001)
	C. gigas	48-hr LC$_{50}$	520–579	
	Culex quinquefasciatus	48-hr LC$_{50}$	127.8	
	C. quinquefasciatus	Effect on reproduction	127.8	
	C. quinquefasciatus	100% mortality	300	
Tiamulin	*D. magna*	48-hr EC$_{50}$	40–67	Boxall et al. (2000)
	Unspecified fish	96-hr LC$_{50}$	5.2	
	Unspecified algae	96-hr EC$_{50}$	>0.62	
	M. aeruginosa	7-d EC$_{50}$	0.003	Halling-Sørensen (2000)
	S. capricornutum	72-hr EC$_{50}$	0.165	
Triclabendazole	Unspecified algae	72-hr EC$_{50}$	45	Lewis and Bardon (1998)
	D. magna	48-hr EC$_{50}$	133	
	Unspecified fish	96-hr EC$_{50}$	117	
Trimethoprim	*M. aeruginosa*	EC$_{50}$	112	Holten Lützhøft et al. (1999)
	S. capricornutum	EC$_{50}$	130	
	R. salina	EC$_{50}$	16	
Tylosin	*D. magna*	48-hr EC$_{50}$	680	Wollenberger et al. (2000)
	M. aeruginosa	7-d EC$_{50}$	0.034	Halling-Sørensen (2000)
	S. capricornutum	72-hr EC$_{50}$	1.38	
Valnemulin	*D. magna*	48-hr EC$_{50}$	44.7	Boxall et al. (2000)
	Fish (yellowtail)		>15 mg kg^{-1} d^{-1}	
	Unspecified aerobic microorganisms	28-d NOEC	>2	

LOEC: lowest observed effect concentration; LC$_{50}$: concentration causing 50% mortality; EC$_{50}$: concentration causing 50% effect; NOEC: no observed effect concentration; LC$_{100}$: concentration causing 100% mortality; EC$_{100}$: concentration causing 100% effect.

Table 9. Terrestrial ecotoxicity data for a range of veterinary medicines.

Compound	Test organism	Toxic effect	Concentration (mg kg^{-1})	Reference
Abamectin	*Onthophagus binodis*	Effect on reproduction and mortality	Dose 200 µg kg^{-1} body weight	Ridsdill-Smith (1993)
	Onthophagus ferox	Effect on reproduction and mortality	Dose 200 µg kg^{-1} body weight	VICH (unpublished)
Apramycin	Earthworm	NOEC	>100	
	Microbes	MIC or NOEC	0.100	
	Unspecified plant	NOEC	160	Elanco (2000)
	Bobwhite quail	14-d oral LD$_{50}$	1669	
	Bobwhite quail	5-d LD$_{50}$ (dietary)	>5000	
	Mallard duck	5-d LD$_{50}$ (dietary)	>5000	
	Azobacter chroococcum	Inhibition	0.1	
	Anabaena flosaquae	Inhibition	0.1	
	R. leguminosarum	Inhibition	0.1	
	R. japonicum	Inhibition	1–10	
Amitraz	Earthworm	14-d LC$_{50}$	1000	Lewis and Bardon (1998)
Amprolium	Nitrification rate of soil	NOEC	>3.06	Warman (1980)
Aureomycin	Nitrification rate of soil	NOEC	>0.34	Warman (1980)
Bacitracin	Microbes	MIC or NOEC	10	VICH (unpublished)
Benzimidazole	Earthworm	—	>1000	Greiner and Ronnefarth (2001)
Ceftiofur	Dung beetle	—	>10	VICH (unpublished)
Chlortetracycline	Microbes	MIC or NOEC	0.250	VICH (unpublished)
	Soil respiration rate	NOEC	>0.6	Warman and Thomas (1981)
Clorsulon	Microbes	MIC or NOEC	2	VICH (unpublished)
Cypermethrin	Collembola	—	Nontoxic	Tomlin (1997)

Table 9. (Continued).

Compound	Test organism	Toxic effect	Concentration (mg kg^{-1})	Reference
Cypermethrin	Bees	24-hr LD$_{50}$	Highly toxic in laboratory tests: 0.035 µg/bee (oral); 0.02 µg/bee (topical)	Tomlin (1997)
Cyromazine	Mallard duck	14-d LD$_{50}$	>2510	USEPA (2001)
	Mallard duck	8-d LC$_{50}$	>5620	
	Honeybee	48-hr LD$_{50}$	>25 µg/bee	
	Northern bobwhite	14-d LD$_{50}$	1785	
	Northern bobwhite	8-d LC$_{50}$	>5620	
	Earthworm	14-d LC$_{50}$	1000	Lewis and Bardon (1998)
Deltamethrin	Bees	LD$_{50}$ (oral)	79 ng/bee	Tomlin (1997)
	Bees	LD$_{50}$ (contact)	51 ng/bee	
	Earthworms	14-d LC$_{50}$	28.6	
	Mallard duck	8-d LC$_{50}$	>4640	USEPA (2001)
	Honeybee	48-hr LD$_{50}$ (topical exposure)	0.067 µg/bee	
	Honeybee	24-hr LD$_{50}$ (dermal exposure)	+0.186	
	Honeybee	1-d LOEL (dermal exposure)	+0.11 µg/bee	
	Northern bobwhite	14-d LD$_{50}$	>2250	
	Northern bobwhite	8-d LC$_{50}$	>10000	
Diazinon	*Lumbricus terrestris*	48-hr LC$_{50}$ (aqueous exposure)	0.0258	Reported in Larkin and Tjeerdema (2000)
	Saprotrophic isopods	LC$_{50}$ (ingestion)	74.15	Vink et al., (1995)
	Saprotrophic isopods	LC$_{50}$	3.03	
	Honeybee	Single dose LD$_{50}$	0.00045 mg/individual	Reported in Larkin and Tjeerdema (2000)

Table 9. (Continued).

Compound	Test organism	Toxic effect	Concentration (mg kg^{-1})	Reference
Dichlorvos	Bees	Oral LD$_{50}$	0.29 µg/bee	Tomlin (1997)
Doramectin	Earthworms	NOEC	2	VICH (unpublished)
	Microbes	MIC or NOEC	40	
	Plants	NOEC	1.6	
	Onthophagus gazella	LC$_{90}$	38.3	Taylor (1999)
	O. gazella	LC$_{50}$	12.5	
	Haematobia irritans	LC$_{90}$	3	
Efrotomycin	Earthworms	NOEC	1000	VICH (unpublished)
	Microbes	MIC or NOEC	20	
	Plants	NOEC	0.40	
Enrofloxacin	Pseudomonas putida	EC$_{50}$	0.0037	Bayer (1997)
Eprinomectin	Earthworm	NOEC	295	VICH (unpublished)
	Plants	MIC or NOEC	1000	
Fenbendazole	Earthworms	NOEC	56	VICH (unpublished)
	Microbes	MIC or NOEC	1000	
	Plants	NOEC	36	
Florfenicol	Microbes	MIC or NOEC	0.4	VICH (unpublished)
Halofuginone	Microbes	MIC or NOEC	200	VICH (unpublished)
	Plants	NOEC	24	
Halothane	Vicia faba (seedling)	Cell mitotic abnormalities over 0.17 d	1%–2%	USEPA (2001)
	V. faba (seedling)	Cell mitotic abnormalities over 0.33 d	1%–2%	
	Alleim cepa	60%–83% reduction in root mitotic rate over 0.33 d	0.5%–2%	

Table 9. (Continued).

Compound	Test organism	Toxic effect	Concentration (mg kg^{-1})	Reference
Ivermectin	Earthworms	NOEC	12	VICH (unpublished)
	Eiseniia foetida	28-d LC$_{50}$	18–100	Halley et al. (1989)
	Plants	NOEC	0.56	VICH (unpublished)
	N. cornicina	Behavior	0.125	Gover and Strong (1996)
	N. cornicina	47% mortality over 7 d (dung)	0.125	
	N. cornicina	77% mortality over 7 d (dung)	0.25	
	N. cornicina	87% mortality over 7 d (dung)	0.5	
	N. cornicina	100% mortality over 7 d (dung)	1	
	N. cornicina	LC$_{50}$ (dung)	0.139	
	Scatophaga stercoraria (larvae)	24-hr EC$_{50}$	0.051	Strong and James (1993)
	S. stercoraria (larvae)	48-hr EC$_{50}$	0.036	
	S. stercoraria (adults)	Developmental abnormalities	0.0005	
	S. stercoraria	50% reduction in pupariation	0.015	Strong and James (1993)
	S. stercoraria	50% reduction in emergence	0.001	Strong and James (1993)
Laidlomycin	Microbes	MIC or NOEC	0.4	VICH (unpublished)
	Plants	NOEC	0.16	
Lasalocid	Microbes	MIC or NOEC	0.20	VICH (unpublished)
	Plants	NOEC	2.0	

Table 9. (Continued).

Compound	Test organism	Toxic effect	Concentration (mg kg^{-1})	Reference
Lincomycin	Earthworms	NOEC	1000	VICH (unpublished)
	Microbes	MIC or NOEC	0.78	USEPA (2001)
	Phaseolus vulgaris (seedling)	Physiological damage to organelle over 1.17 d	100 µg mL^{-1}	
	P. vulgaris (seedling)	12% reduction to leaf chlorophyll over 0.12 d	100 µg mL^{-1}	
	P. vulgaris (seedling)	39% reduction to leaf chlorophyll over 0.25 d	100 µg mL^{-1}	
	P. vulgaris (seedling)	23% reduction to leaf chlorophyll over 0.79 d	100 µg mL^{-1}	
	P. vulgaris (seedling)	39% reduction to leaf chlorophyll over 0.92 d	100 µg mL^{-1}	
	P. vulgaris (seedling)	61% reduction to leaf chlorophyll over 2 d	100 µg mL^{-1}	
	P. vulgaris (seedling)	61% reduction to leaf chlorophyll over 2.67 d	100 µg mL^{-1}	
	P. vulgaris (seedling)	43% reduction in leaf photosynthesis over 0.25 d	100 µg mL^{-1}	
	P. vulgaris (seedling)	43% reduction in leaf photosynthesis over 0.92 d	100 µg mL^{-1}	
	P. vulgaris (seedling)	86% reduction in leaf photosynthesis over 1 d	100 µg mL^{-1}	
	P. vulgaris (seedling)	92% reduction in leaf photosynthesis over 2.67 d	100 µg mL^{-1}	
	P. vulgaris (seedling)	Leaf pigmentation over 2.67 d	100 µg mL^{-1}	
Maduramicin	Microbes	MIC or NOEC	0.25	VICH (unpublished)
	Plants	NOEC	0.1	

Table 9. (Continued).

Compound	Test organism	Toxic effect	Concentration (mg kg^{-1})	Reference
Melengestrol acetate	Earthworms	NOEC	1.8	VICH (unpublished)
	Plants	NOEC	2	VICH (unpublished)
Monensin	Earthworms	NOEC	10	VICH (unpublished)
	Plants	MIC or NOEC	0.15	VICH (unpublished)
	Bobwhite quail	14-d oral LD$_{50}$	85.7	Elanco (1998)
	Bobwhite quail	5-d LD$_{50}$ (dietary)	1090	
	Mallard duck	5-d LD$_{50}$ (dietary)	>5000	
	Earthworm	14-d LD$_{50}$	>100	
Morantel	Microbes	MIC or NOEC	50	VICH (unpublished)
Narasin	Earthworms	NOEC	0.5	VICH (unpublished)
Narasin	Microbes	MIC or NOEC	0.1	VICH (unpublished)
	Plant	NOEC	0.15	
Oxfendazole	Earthworm	NOEC	971	VICH (unpublished)
	Microbes	MIC or NOEC	9	
	Plants	NOEC	0.9	
Oxytetracycline	Mallard duck	8-d LC$_{50}$	>5620 ppm	USEPA (2001)
	Northern bobwhite	8-d LC$_{50}$	>5620 ppm	
	Northern bobwhite	14-d LD$_{50}$	>2000	
	Folsomia. fimetaria	LC$_{50}$	>5000 mg kg^{-1}	Baguer et al. (2000)
	F. fimetaria	EC$_{50}$ reproduction	>5000 mg kg^{-1}	
	Enchytraeus crypticus	LC$_{50}$	>5000 mg kg^{-1}	
	E. crypticus	EC$_{50}$ reproduction	2701 mg kg^{-1}	
	Aporrectodea caliginosa	LC$_{50}$	>5000 mg kg^{-1}	
	A. caliginosa	EC$_{50}$ reproduction	4420 mg kg^{-1}	
	A. caliginosa	EC$_{50}$ growth	>5000 mg kg^{-1}	
	A. caliginosa	EC$_{50}$ hatchability	>5000 mg kg^{-1}	

Table 9. (Continued).

Compound	Test organism	Toxic effect	Concentration (mg kg^{-1})	Reference
Pirlimycin	Earthworm	NOEC	1000	VICH (unpublished)
	Microbes	MIC or NOEC	0.13	
	Plants	NOEC	0.4	
Procaine penicillin	*Lactus sativa*	Effect on root mitotic rate and hypocotyl size over 2.5 d	0.5%	USEPA (2001)
Salinomycin	Microbes	MIC or NOEC	0.78	VICH (unpublished)
	Plants	NOEC	0.4	
Sarafloxicin	Earthworm	NOEC	1000	VICH (unpublished)
	Microbes	MIC or NOEC	0.03	
	Plants	NOEC	1.3	
Sendramicin	Microbes	MIC or NOEC	100	VICH (unpublished)
	Plants	NOEC	0.31	
Sulfadiazine	*Lupinus albus*	13% reduction in root size over 1 d	100 ppm	USEPA (2001)
Sulfadimethoxine	*Amaranthus retroflexus*	Development	<300 mg L^{-1}	Migliore et al. (1997b)
	Plantago major	Development	<300 mg L^{-1}	
	Rumex acetosella	Development	<300 mg L^{-1}	
	Paniceum miliaceum	Development	<300 mg L^{-1}	Migliore et al. (1995)
	Pisum sativum	Development	<300 mg L^{-1}	
	Zea mays	Development	<300 mg L^{-1}	
	Hordeum disthicum	Development and growth	<300 mg L^{-1}	Migliore et al. (1996)
Teflubenzuron	Bees	Nontoxic at recommended rates	—	Tomlin (1997)
	Other	Low toxicity to predatory arthropods	—	

Table 9. (Continued).

Compound	Test organism	Toxic effect	Concentration (mg kg^{-1})	Reference
Tiamulin	Wheat	Plant vigor/germination	No effect	Boxall et al. (2000)
	Lettuce	Plant vigor/germination	No effect	
	Microbes	MIC or NOEC	500	VICH (unpublished)
Tylosin	F. fimetaria	LC$_{50}$	>5000	Baguer et al. (2000)
	F. fimetaria	EC$_{50}$ reproduction	2520	
	E. crypticus	LC$_{50}$	3381	
	E. crypticus	EC$_{50}$ reproduction	3109	
	A. caliginosa	LC$_{50}$	>5000	
	A. caliginosa	EC$_{50}$ reproduction	4530	
	A. caliginosa	EC$_{50}$ growth	>5000	
	A. caliginosa	EC$_{50}$ hatchability	4823	
	Bobwhite quail	5-d LD$_{50}$ (dietary)	4820	Elanco (2000)
	Mallard duck	5-d LD$_{50}$ (dietary)	4710	
	Earthworm	28-d LD$_{50}$	918	
	Chaetomium globosum		>1000	
	Aspergillus flavus	Inhibition	>1000	
	Comamonas acidvorans	Inhibition	250	
	Azobacter chroococcum	Inhibition	5	
Virginiamycin	Microbes	MIC or NOEC	10	VICH (unpublished)

In current risk assessment approaches (e.g., biocides, pesticides, and industrial chemicals), a factor of 10 is typically used to account for differences between acute and chronic endpoints, and the data given in Table 10 support its use for veterinary medicines.

Ethinyl Estradiol Both long-term and short-term toxicity studies have been performed on ethinyl estradiol (Kopf 1995; Schweinfurth et al. 1996). These studies indicate that ethinyl estradiol is nontoxic to microbes and that toxicity to daphnids and algae is generally in the low mg L^{-1} range. However, long-term studies into the toxicity of ethinyl estradiol to fish indicated that fish growth is reduced in larvae at concentrations exceeding 100 ng L^{-1}. Moreover, histological changes in the kidney and liver of larvae and juvenile fish have been reported at concentrations as low as 10 ng L^{-1}.

B. Terrestrial Effects

Data were available on the toxicity of 45 chemicals used in veterinary medicines to terrestrial organisms (see Table 9). These tests covered a range of species, including microbes, plants, earthworms, and insects, and a range of endpoints. Five main classes of product have been studied: the endectocides, sheep dip chemicals, antibacterial agents, anticoccidials, and performance enhancers.

Antiparasitics For chemicals used in sheep dip formulations, the practice of applying spent sheep dip to land as a means of disposal may have implications with regards to toxicity to sensitive terrestrial ecosystems. Acute toxicity studies have shown diazinon to be highly toxic to earthworms [48-hr LC_{50}, 25.8 µg L^{-1} (aqueous exposure route)] (Larkin and Tjeerdema 2000). In addition, the toxicity of diazinon to saprophytic isopods has been shown to be dependent on the route of exposure. In studies where substrate exposure was assessed using contaminated sand and dietary exposure was evaluated by feeding organisms contami-

Table 10. Acute and chronic toxicity values for a range of antibacterial agents to *Daphnia magna*.

	Acute		Chronic		
	LOEC (mg L^{-1})	EC_{50} (mg L^{-1})	EC_{50} (mg L^{-1})	NOEC (mg L^{-1})	Ratio
Metronidazole	1000			250	4
Oxolinic acid		4.6		0.75	6
Oxytetracycline	100		46.2		2.2
Streptomycin		487		32	15
Sulfadiazine		221	13.7		16
Tetracycline	340		44.8		7.6
Tiamulin		32	5.4		5.9
Tylosin		483		90	5.4

Source: Data from Wollenberger et al. (2000).

nated leaf material, the former was found to be far more lethal (Vink et al. 1995). In laboratory studies, cypermethrin and diazinon were shown to strongly affect honeybees, with lethal topical doses of 0.02 and 0.45 μg per bee, respectively, reported (Larkin and Tjeerdema 2000; Tomlin 1997). Apart from diazinon, very few data were available on the toxicity to terrestrial invertebrates of chemicals used in sheep dip preparations.

The avermectins are powerful insecticides that are thought to exhibit their effect on the γ-aminobutyric acid-mediated neuromuscular synapse, with chloride channels appearing to be particularly sensitive (Turner and Schaeffer 1989). Exposure to avermectins can elicit a number of responses, including adult and larval mortality, an effect on feeding, disruption of water balance, a reduction in growth rate, interference with molting, inhibition of metamorphosis and/or pupation, prevention of adult emergence, disruption of mating, and interference with egg production and oviposition (Strong 1993; Strong and Brown 1987). As a consequence, dung from animals treated with avermectins may not support the development of either target (e.g., *Haemotobia irritans, Musca autumnnalis, Musca domestica, Musca vetustissimia*) or nontarget (e.g., sphaerocerids, muscids, sepsids, coleopterans) insects (Madsen et al. 1990; Miller et al. 1981; Ridsdill-Smith 1988; Schmidt 1983; Sommer et al. 1992, 1993; Strong and Brown 1987; Strong and James 1993; Wall and Strong 1987). The toxicity of avermectins to dung insect populations may be associated with a retardation in the rate of breakdown of pats. For example, pats containing ivermectin have been shown to be intact after 340 d, whereas untreated pats were largely degraded within 80 d (Floate 1998).

The effects on other invertebrates have not been extensively investigated, although investigations with annelids demonstrated no effect on population density (Wall and Strong 1987). The possible indirect effects of avermectin-contaminated dung on vertebrate populations have also been highlighted (McCracken 1993). Their use may result in depletion of the quantity and quality of vertebrate food resources, which may be particularly critical during the breeding season or when young animals are foraging and fending for themselves.

Antibacterial Agents, Anticoccidials, and Growth Promoters Summary data were available on the toxicity of antibacterial agents, anticoccidials, and performance enhancers to earthworms, microbes, and plants (VICH 2000). For the antibacterial agents, microbes were the most sensitive test species with minimum inhibitory concentrations (MICs) or no observed effect concentrations (NOECs) ranging from 100 (apramycin) to 500,000 (tiamulin) μg kg^{-1}. For the anticoccidials, plants and microbes were the most sensitive with microbial inhibition concentrations or NOECs ranging from 100 (narasin) to 200,000 (halofuginone) μg kg^{-1}. For the growth promoters, plants are shown to be sensitive to monensin (NOEC, 150 μg kg^{-1}) and microbes to lasalocid sodium (NOEC, 2000 μg kg^{-1}).

C. Oestrogenic Activity

A number of compounds and environmental effluents have been associated with potential reproductive and developmental abnormalities in fish. For example, hermaphrodite fish have been observed in rivers below sewage treatment plants

(Harries et al. 1997; Purdom et al. 1994). Four chemicals used in veterinary products have been identified as exhibiting endocrine-disrupting properties, namely estradiol, ethinyl estradiol, diazinon, and permethrin (Environment Agency 1998). For example, ethinyl estradiol has been shown to reduce egg deposition in adult fish at concentrations of 10 ng L^{-1}, and a 9-mon study with fish resulted in a NOEC for reproduction of 1 ng L^{-1} (Schweinfurth et al. 1996).

Limited data are available on the fate of estradiol and ethinyl estradiol used as veterinary medicines (Arcand-Hoy et al. 1998); consequently, it is difficult to assess the importance of their use in veterinary products in terms of estrogenic effects on the aquatic environment. Data are available, however, on endogenous estrogens (Shore et al. 1988) that demonstrate that these compounds can be transported from poultry farms via agricultural runoff to rivers and streams. The reported concentrations of the endogenous estrogens in manure were 66 ng g^{-1}. Concentrations in water collected from four streams ranged from 0.8 to 10.4 ng L^{-1}.

It is possible that other veterinary medicines may cause endocrine disruption in ecosystems. However, because of the lack of appropriate screening methods, ecologically significant changes in reproductive function resulting from endocrine-disruptive effects of chemicals are not routinely detected.

VIII. Recommendations for Further Work

This article has reviewed the data available in the public domain on the usage, pathways to the environment, fate, and effects of veterinary medicines. Although there is a large body of data available, there are clearly a number of gaps in the data and in our understanding of the impacts of veterinary medicines on the environment. On the basis of the review, a number of priority areas for future work were identified.

1. Usage data are unavailable for many groups of veterinary medicines and for many countries, which makes it difficult to establish whether these substances pose a risk to the environment. It is therefore recommended that usage information be obtained for these groups, including antiseptics, steroids, diuretics, cardiovascular and respiratory treatments, locomotor treatments, and immunological products. Better usage data will enable properly designed and science-based effective intervention and mitigation strategies.
2. From the information available, it appears that inputs from aquaculture and herd or flock treatments are probably the most significant in terms of environmental exposure. The relative significance of other routes of entry to the environment from livestock treatments, such as washoff following topical treatment, farm yard runoff, and aerial emissions, have not generally been considered. These routes should therefore be assessed and incorporated into risk assessment models if appropriate. Other routes of release should also be considered. For example, the significance of exposure to the environment from the disposal of used containers or from discharge from manufacturing

sites should be investigated further. In addition, substances may be released to the environment as a result of off-label use and poor slurry management practice. The significance of these exposure routes is currently unknown.

3. Monitoring data are available for a range of veterinary medicines in soil, sediment, surface waters, and groundwaters. The studies have generally focused on the sheep dip chemicals, the anthelmintics, aquaculture treatments, and the antibiotics. Once full datasets are obtained on the usage, properties, and effects of other chemical groups, further targeted monitoring should be performed to determine concentrations in the environment. These data could then be used along with the existing data to evaluate current risk assessment exposure models.

4. With the exception of a few groups, e.g., anthelmintics and sheep dip chemicals, ecotoxicity studies have only been performed on a limited number of species (particularly terrestrial species). For groups that are identified as potentially being of high environmental risk, more extensive studies should be conducted to establish species sensitivity distributions and likely impacts on the agricultural landscape.

5. A number of hormones are used as veterinary medicines. However, limited information is available on the endocrine-disrupting potential of these substances. A more detailed assessment of the endocrine-disrupting potential of hormones (and other substances) used as veterinary medicines should therefore be performed. This information should be combined with information on likely releases to the environment to assess the potential for veterinary medicines to disrupt endocrine systems in the environment.

6. Although information was available on the direct effects of a range of veterinary medicines on aquatic and terrestrial organisms, limited information is available on the indirect effects. The possible indirect effects of veterinary medicines should be identified. For example, concern has been raised over the possible indirect effects of anthelmintics on higher trophic levels (such as bat or bird species) that may result from the loss of dung invertebrates as a food source.

7. Finally, there is a need to establish the potential for veterinary medicines to bioaccumulate. It may be possible to perform these assessments using data on target animals obtained in pharmacodynamic/pharmacokinetic studies performed by manufacturers as part of the current regulatory process.

By performing studies of this type, it should be possible to fully establish the risks posed to the environment by veterinary medicines and, where appropriate, introduce strategies to manage these risks.

Summary

The impact of veterinary medicines on the environment will depend on a number of factors including physicochemical properties, amount used and method of administration, treatment type and dose, animal husbandry practices, manure storage and handling practices, metabolism within the animal, and degradation rates in manure and slurry. Once released to the environment, other factors such as soil type, climate, and ecotoxicity also determine the environmental impact of the compound.

The importance of individual routes into the environment for different types of veterinary medicines varies according to the type of treatment and livestock category. Treatments used in aquaculture have a high potential to reach the aquatic environment. The main routes of entry to the terrestrial environment are from the use of veterinary medicines in intensively reared livestock, via the application of slurry and manure to land, and by the use of veterinary medicines in pasture-reared animals where pharmaceutical residues are excreted directly into the environment. Veterinary medicines applied to land via spreading of slurry may also enter the aquatic environment indirectly via surface runoff or leaching to groundwater. It is likely that topical treatments have greater potential to be released to the environment than treatments administered orally or by injection. Inputs from the manufacturing process, companion animal treatments, and disposal are likely to be minimal in comparison.

Monitoring studies demonstrate that veterinary medicines do enter the environment, with sheep dip chemicals, antibiotics, sealice treatments, and anthelmintics being measured in soils, groundwater, surface waters, sediment, or biota. Maximum concentrations vary across chemical classes, with very high concentrations being reported for the sheep dip chemicals.

The degree to which veterinary medicines may adsorb to particulates varies widely. Partition coefficients (K_d) range from low (0.61 L kg^{-1}) to high (6000 L kg^{-1}). The variation in partitioning for many of the compounds in different soils was significant (up to a factor of 30), but these differences could be not be explained by normalization to the organic carbon content of the soils. Thus, to arrive at a realistic assessment of the availability of veterinary medicines for transport through the soil and uptake into soil organisms, the K_{oc} (which is used in many of the exposure models) may not be an appropriate measure. Transport of particle-associated substances from soil to surface waters has also been demonstrated. Veterinary medicines can persist in soils for days to years, and half-lives are influenced by a range of factors including temperature, pH, and the presence of manure. The persistence of major groups of veterinary medicines in soil, manure, slurry, and water varies across and within classes.

Ecotoxicity data were available for a wide range of veterinary medicines. The acute and chronic effects of avermectins and sheep dip chemicals on aquatic organisms are well documented, and these substances are known to be toxic to many organisms at low concentrations (ng L^{-1} to µg L^{-1}). Concerns have also been raised about the possibility of indirect effects of these substances on predatory species (e.g., birds and bats). Data for other groups indicate that toxicity values are generally in the mg L^{-1} range. For the antibiotics, toxicity is greater for certain species of algae and marine bacteria. Generally, toxicity values for antibacterial agents were significantly higher than reported environmental concentrations. However, because of a lack of appropriate toxicity data, it is difficult to assess the environmental significance of these observations with regard to subtle long-term effects.

Appendix A. CAS numbers and systematic names of veterinary medicines reported in this paper.

Compound	CAS	Systematic name
Abamectin	71751-41-2	5-*O*-Demethyl avermectin A1a and 5-*O*-demethyl-25-de(1-methylpropyl)-25-methylethyl) avermectin A1a
Altrenogest	850-52-2	17-Allyl-17β-hydroxyestra-4,9,11-trien-3-one
Aminosidine	7542-37-2	2-Amino-2-deoxy-α-D-glucopyranosyl-(1 to 4)-*O*-[*O*-2,6-diamino-2,6-dideoxy-β-L-idopyranosyl-(1 to 3)-β-D-ribofuranosyl-(1 to 5)]-2-deoxy-, sulfate (salt)
Amitraz	33089-61-1	*N'*-(2,4-Dimethylphenyl)-*N*-[[(2,4-dimethylphenyl)imino]methyl]-*N*-methylmethanimidamide
Amoxicillin	61336-70-7	(2*S*,5*R*,6*R*)-6[(*R*-2-amino-2-(4-hydroxyphenyl)acetamido]-3,3-dimethyl-7-oxo-4-thia-1-azabicyclo-2-carboxylic acid
Ampicillin	69-53-4	2*S*-[2a,5a,6b(*S**)]]-6[(Aminophenylacetyl)amino]-3,3-dimethyl-7-oxo-4-thia-1-azabicyclo(3,2,0)heptane-2-carboxylic acid
Amprolium	137-88-2	4-Acetamido-2-ethoxy benzoic acid methylester
Apramycin	37321-09-8	D-Steptamine, *O*-4-amino-4-deoxy-α-D-glucopyranosyl-(→8)-*O*-(8*R*)-2-amino-2,3,7-trideocy-7-(methylamino)-D-glycero-α-D-allo-octodialdo-1,5:8,4-dipyranosyl-(1→4)-2-deoxy, sulfate
Avermectin B1a	73989-17-0	5-*O*-Demethyl avermectin A1a
Azamethiphos	35575-96-3	*S*-(6-Chloro-oxazolo[4,5-b]pyridin-2(3*H*)-on-3-ylmethyl) *O*,*O*-dimethyl phosphorothioate
Bacitracin	1405-87-4	Comprises antimicrobial polypeptides yielding the following amino acids on hydrolysis: L-cysteine; D-glutamic acid; L-histidine; L-isoleucine; L-lysine; D-ornithine; D-phenylalanine; DL-aspartic acid
Benzyl alcohol	100-51-6	(Hydroxymethyl)benzene
Ceftiofur	80370-57-6	(6*R*,7*R*)-7-[2-(2-Amino-4-tiazoyl)-2-((Z)-methoxyamino)acetamido]-(20-furylthiomethyl)-8-oxo-5-thia-azabicyclo-4.3.0)oct-2-ene-2-carboxylic acid
Chloramphenicol	56-75-7	D-(−)-*threo*-2,2-Dichloro-*N*-[β-hydroxy-α-(hydroxymethyl)-*p*-nitrophenethyl]acetamide
Chlorfenvinphos	470-90-6	Diethyl-*O*-(2-chloro-1-(2′,4′-dichlorophenyl)vinyl) phosphate
Chlorhexidine	55-56-1	Hexamethylene-bis(5-(4-chlorophenyl)biguanide)

Appendix A. (*Continued*)

Chlortetracycline	64-72-2	7-Chloro-4-(dimethylamino)-1,4,4a,5,5a,6,11,12a-octahydro-3,6,10,12,12a-pentahydroxy-6-methyl-1,11-dioxo-,(4S-(4α,4α,5α,6β,12α))-naphthacenecarboxamide
Cinoxacin	28657-80-9	1-Ethyl-1,4-dihydro-4-oxo(1,3)dioxolo(4,5-g)cinnoline-3-carboxylic acid
Ciprofloxacin	85721-33-1	1-Cyclopropyl-6-fluoro-1,4-dihydro-4-oxo-7-(1-piperazinyl)-3-quinolinecarboxylic acid
Clyndamycin	18323-44-9	Methyl 7-chloro-6,7,8-trideoxy-6-(((1-methyl-4-propyl-2-pyrrolidinyl)carbon yl) amino)-1-thio-, (2S-*trans*)-, hydrochloride monohydrate
Coumaphos	56-72-4	*O*-(3-Chloro-4-methyl-2-oxo-2*H*-1-benzopyran-7-yl) *O,O*-diethyl phosphorothioate
Cyromazine	66215-27-8	Cyclopropyl-1,3,5-triazine-2,4,6-triamine
Cypermethrin	52315-07-8	(*RS*)-α-Cyano-3-phenoxybenzyl (1*RS*,3*RS*; 1*RS*,3*SR*)-3-(2,2-dichlorovinyl)-2,2-dimethylcyclopropanecarboxylate
Danofloxacin	112398-08-0	(1*S*)-1-Cyclopropyl-6-fluoro-1,4-dihydro-7-(5-methyl-2,5-diazabicyclo[2.2.1]hept-2-yl)-4-oxo-3-quinolinecarboxylic acid
Delmadinone	15262-77-8	6-Chloro-17-hydroxypregna-1,4,6-triene-3,20-dione
Deltamethrin	52918-63-5	(*S*)-α-Cyano-3-phenoxybenzyl (1*R*,3*R*)-3-(2,2,-dibromovinyl)-2,2-dimethyl cyclopropanecarboxylate
Diazinon	333-41-51	*O,O*-Diethyl *O*-2-isopropyl-6-methylpyrimidin-4-yl phosphorothioate
Dichlorvos	62-73-7	2,2-Dichlorovinyl dimethyl phosphate
Dihydrostreptomycin	128-46-1	*O*-2-Deoxy-2-(methylamino)α-*L*-glucopyranosyl-(1-2)-*O*-5-deoxy-3-*C*-(hydroxymethyl)-α-L-lyxofuranosyl-(1-4)*N*,*N*′-diaminido-D-streptamine
Dimethicone	9016-00-6	Dimethyl silicone
Dimetridazole	551-92-8	1,2-Dimethyl-5-nitro-1*H*-imidazole
Doramectin	117704-25-3	25-Cyclohexyl-5-*O*-demethyl-25-de (1-methyl propyl)avermectin-1a
Doxycycline	564-25-0	4-(Dimethylamino)-1,4,4a,5,5a,6,11,12a-octahydro-3,5,10,12,12a-pentahydroxy-6-methyl-1,11-dioxo-2-naphthacenecarboxamide monohydrate
Efrotomycin	56592-32-6	Efrotomycin
Emamectin benzoate	137512-74-4	4″-Epimethylamino-4″-deoxy avermectin B1a and B1b
Enoxacin	74011-58-8	1-Ethyl-6-fluoro-1,4-dihydro-4-oxo-7-(1-piperazinyl)-1,8-naphthyridine-3-carboxylic acid
Enrofloxacin	93106-60-6	1-Cyclopropyl-7-(4-ethyl-1-piperazinyl)-6-fluoro-1,4-dihydro-4-oxo-3-quinolinecarboxylic acid

Appendix A. (*Continued*)

Eprinomectin	159628-36-1	4″-epi-Acetylamino-4″-deoxy-avermectin B1
Erythromycin	114-07-8	(2R,3S,4S,5R,6R,8R,10R,11R,12S,13R)-5-(3,4,6-Trideoxy-3-dimethylamino-β-D-xylo-hexopyranosyloxy)-3-(2,6-dideoxy-3-methyl-3-O-methyl-α-L-*ribo*-hexopyranosyloxy)-13-ethyl-6,11,12,-trihydroxy-2,4,6,8,10,12-hexamethyl-9-oxotridecan-13-olide
Estradiol benzoate	50-50-0	17β-Estradiol-3-benzoate
Ethenyl estradiol	57-63-6	17α-Ethynyl-3-hydroxy-1,3,5(10)-estratrien-17β-ol
Fenbendazole	43210-67-9	[5-(Phenylthio)-1*H*-benzimidazol-2-yl]carbamic acid methyl ester
Fenchlorphos	299-84-3	*O,O*-Dimethyl *O*-(2,4,5-trichlorophenyl) phosphorothioate
Florfenicol	76639-94-6	[R—(R1,T)]-2,2-dichloro-*N*-{fluoromethyl:-2-hydroxy-2-[4-(methylsulfonyl:phenyl] ethyl}acetamide
Flumethrin	69770-45-2	Cyano(4-fluoro-3-phenoxyphenyl)methyl 3-[2-chloro-2-(4-chlorophenyl)ethenyl]-2,2-dimethylcyclopropanecarboxylate
Formosulphathiazole	72-14-0	*N*1-(2-Thiazolyl)sulfanilamide
Furazolidone	67-45-8	3-[[(5-Nitro-2-furanyl)methylene]amino]-; 3-(5-nitro-2-furyl)methylene amino-2-oxazolidinone
Griseofulvin	126-07-8	7-Chloro-4,6-dimethoxycoumaran-3-one-2-spiro-1-(2′-methoxy-6′-methyl cyclohex-2′-en-4′-one)
Halofuginone	55837-20-2	*trans*-7-Bromo-3[3-(3-hydroxy-2-piperidyl)acetonyl]-6-chloro-4(3*H*)quinazolinone
Halothane	151-67-7	2-Bromo-2-chloro-1,1,1-trifluoroethane
Hydrogen peroxide	7722-84-1	Dihydrogen dioxide
Isoflurane	26675-46-7	1-Chloro-2,2,2-trifluoroethyl difluoromethyl ether
Ivermectin	70288-86-7	5-0-Demethyl-22, 23-dihydroavermectin A_{1a} (component B_{1a}) and 5-0-demethyl-25-de (1-methylpropyl)-22, 23-dihydro-25-(1-methylethyl)avermectin A_{1b} (component B_{1b})
Laidlomycin	56283-74-0	16-Deethyl-3-*O*-demethyl-16-methyl-3-*O*-(1-oxopropyl)monensin
Lasalocid	25999-31-9	6-(7-(5-Ethyl-5-(5-ethyltetrahydro-5-hydroxy-6-methyl-2*H*-pyran-2-yl)tetrahydro-3-methyl-2-furyl)-4-hydroxy-3,5-dimethyl-6-oxononyl)-, (-)-2,3-cresotic acid
Levamisole	14769-73-4	*S*(−)-6-Phenyl-2,3,5,6-tetrahydroimidazo [2,1-b]thiazole

Appendix A. (*Continued*)

Lido/lignocaine	137-58-6	2-(Diethylamino)-*N*-(2,6-dimethylphenyl)acetamide
Lincomycin	154-21-2	Methyl 6,8-dideoxy-6-(1-methyl-4-propyl-2-pyrrolidinecarboxamido)-1-thio-D-erythro-α-D-galactooctopyranoside
Lomefloxacin	98079-51-7	1-Ethyl-6,8-difluoro-1,4-dihydro-7-(3-methyl-1-piperazinyl)-4-oxo-3-quinolinecarboxylic acid
Maduramicin	84878-61-5	Maduramicin
Medroxyprogesterone	520-85-4	17a-Hydroxy-6a-methyl-4-pregnene-3,20-dione
Melengestrol acetate	2919-66-6	6-Dehydro-17-hydroxy-6-methyl-16-methylene-progesterone acetate
Meropenem	96036-03-2	3-((5-((Dimethylamino)carbonyl)-3-pyrrolidinyl)thio)-6-(1-hydroxyethyl)-4-methyl-7-oxo-(4*R*-(3(*S**,5*S**),4α,5β,6β(*R**)))-1-azabicyclo(3.2.0)hept-2-ene-2-carboxylic acid
Metamyzole		*N*-(2,3-Dihydro-1,5-dimethyl-3-oxo-2-phenyl-4-pyrazolyl)-*N*-methylamino]methansulfonic acid
Methyltestosterone	58-18-4	17β-Hydroxy-17α-methylandrost-4-en-3-one
Metronidazole	443-48-1	2-Methyl-5-nitro-1-imidazoleethanol
Miconazole	22916-47-8	1-[2-(2,4-Dichlorophenyl)-2-[(2,4-dichlorophenyl)methoxy]ethyl]-1*H*-imidazole
Monensin	17090-79-8	2-[5-Ethyltetrahydro-5-[tetrahydro-3-methyl-5-[tetrahydro- 6-hydroxy-6-(hydroxymethyl)-3,5-dimethyl-2H-pyran-2-yl]-2-furyl-9-hydroxy-β-methoxy-α,γ, 2,8-Tetramethyl- 1,6-dioxaspiro [4.5]decane-7-butyric acid; monensic acid
Morantel	20574-50-9	1,4,5,6-Tetrahydro-1-methyl-2-[2-(3-methyl)ethenyl] pyrimidine
Nalidixic acid	389-08-2	1,4-Dihydro-1-ethyl-7-methyl-4-oxo-1,8-naphthyridine-3-carboxylic acid
Narasin	55134-13-9	4-Methylsalinomycin
Neomycin	1404-04-2	4-*O*-(2,6-Diamino-2,6-dideoxy-α-D-glucopyranosyl)-5-*O*-[3-*O*-(2,6-diamino-2,6-dideoxy-β L-idopyranosyl)-β-D-ribofuranosyl]-2-deoxy-D-streptamine
Nicarbazin	330-95-0	4,4'-Dinitrocarbanilide-2-hydroxy-4,6-dimethyl pyrimidine
Nitroxynil	1689-89-0	4-Hydroxy-3-iodo-5-nitrobenzonitrile
Norfloxacin	70458-96-7	1-Ethyl-6-fluoro-1,4-dihydro-4-oxo-7-(1-piperazinyl)-3-quinolinecarboxylic acid
Ofloxacin	83380-47-6	(±)-9-Fluoro-2,3-dihydro-3-methyl-10-(4-methyl-1-piperazinyl)-7-oxo-7*H*-pyrido[1,2,3-de]-1,4 benzoxazine-6-carboxylic acid

Appendix A. (*Continued*)

Olaquindox	23696-28-8	*N*-(2-Hydroxyethyl)-3-methyl-2-quinoxalinecarboxamide 1,4-dioxide
Oxfendazole	53716-50-0	[5-(Phenylsulfinyl)-1*H*-benzimidazol-2-yl]carbamic acid methyl ester; Synanthic; oxfendazole
Oxolinic acid	14698-29-4	5-Ethyl-5,8-dihydro-8-oxo-1,3-dioxolo (4,5-g)quinoline-7-carboxylic acid
Oxytetracycline	79-57-2	(4*S*,4a*R*,5*S*,5a*R*,6*S*,12a*S*)-4-Dimethylamino-1,4,4a,5,5a,6,11,12a-octahydro-3,5,6,10,12,12a-hexahydroxy-6-methyl-1,11-dioxonaphthacene-2-carboxamide
Pentobarbitone	76-74-4	5-Ethyl-5-(1-methylbutyl)-2,4,6(1*H*,3*H*,5*H*)-pyrimidinetrione
Phenobarbitone	50-06-6	5-Ethyl-5-phenylbarbituric acid
Phenylbutazone	50-33-9	4-Butyl-1,2-diphenyl-3,5-dioxo pyrazolidine
Phosmet	732-11-6	Dimethyl phosphorodithioate *S*-ester with *N*-(mercaptomethyl)phthalimide
Pipemidic acid	51940-44-4	8-Ethyl-5,8-dihydro-5-oxo-2(1-piperazinyl)-pyrido[2,3-d]pyrimidine-6-carboxylic acid (trihydrate)
Pirlimycin	79548-73-5	Pirlimycin
Poloxalene	9003-11-6	Poloxyporopylene-poloxyethylene glycol nonionic polymer
Procaine	59-46-1	4-Aminbenzoic acid-2(diethylanimo)ethyl ester
Procaine benzylpenicillin	61-33-6	4-Thia-1-azabicyclo (3,2,0) hepatne-2-carboxylic acid, 3,3-dimethyl-7-oxo-6-(2-phenylacetamido)
Procaine penicillin	69-57-8	3,3-Dimethyl-7-oxo-6-(2-phenyl-acetamido)-4-thia-1-azabicyclo[3.2.0]heptane-2-carboxylic acid, monosodium salt
Progesterone	57-83-0	(*S*)-Pregn-4-en-3,20-dione
Propetamphos	31218-83-4	1-Methylethyl 3-(((ethylamino)methoxyphosphinothioyl)oxy)-2-butenoate
Pyrantel	15686-83-6	1,4,5,6-Tetrahydro-1-methyl-2-[2-(2-thienyl)ethenyl] pyrimidine
Salinomycin	53003-10-4	(2*R*)-2-[(2*R*,5*S*,6*R*)-6-[(1*S*,2*S*,3*S*,5*R*)-5-[(2*S*,5*S*7*R*,9*S*,10*S*,12*R*,15*R*)-2-[(2*R*,5*R*,6*S*)-5-Ethyltetrahydro-5-hydroxy-6-methylpyran-2-yl]-15-hydroxy-2,10,12-trimethyl-1,6,8-trioxadispiro[4.1.5.3]pentadec-13-en-9-yl]-2-hydroxy-1,3-dimethyl-4-oxoheptyl]tetrahydro-5-methylpyran-2-yl]butyrate
Sarafloxicin	98105-99-8	3-Quinolinecarboxylic acid, 6-fluoro-1-(4-fluorophenyl)-1,4-dihydro-4-oxo-7-(1-piperazinyl), mono-hydrochloride

Appendix A. (Continued)

Spiramycin	8025-81-8	(4R,5S,6S,7R,9R,10R,16R)-(11E,13E)-6-[O-2,6-Dideoxy-3-C-methyl-α-L-ribo-hexopyranosyl-(1→4)-(3,6-dideoxy-3-dimethylamino-β-D-glucopyranosyl)oxy]-7-formylmethyl-4-hydroxy-5-methoxy-9,16-dimethyl]-10-[(2,3,4,6-tetradeoxy-4-dimethylamino-D-erythro-hexopyranosyl)oxy]oxacyclohexadeca-11,13-dien-2-one
Sulfachloropyridazine	80-32-0	4-Amino-N-(6-chloro-3-pyradizinyl)-benzenesulfonamide
Sulfadimethoxine	122-11-2	6-Sulfanilamido-2,4-dimethoxypyrimidine
Sulfamethazine	57-68-1	N-(4,6-Dimethyl-2-pyrimidyl)sulfanilamide
Sulfadiazine	68-35-9	Benzenesulfonamide, 4-amino-N-2-pyrimidinyl
Sulfadimidine	57-68-1	N-(4,6-Dimethyl-2-pyrimidyl)sulfanilamide
Teflubenzuron	83121-18-0	N-(((3,5-Dichloro-2,4-difluorophenyl)amino)carbonyl)-2,6-difluorobenzamide
Tetracycline	60-54-8	(4S-(4α,4aα,5α,6β,12α))-4-(Dimethylamino)-1,4,4a,5,5a,6,11,12a-octahydro-3,6,10,12,12a-pentahydroxy-6-methyl-1,11-dioxo-2-naphthacenecarboxamide
Tiamulin	55297-95-5	14-Desoxy-14-[(2-diethylaminoethyl)mercaptoacetoxy]mutilin hydrogen fumarate
Triclabendazole	68786-66-3	6-Chloro-5-(2,3-dichlorophenoxy)-2-methylthio-benzimidazole
Trimethoprim	738-70-5	5-((3,4,5-Trimethoxyphenyl)methyl)-2,4-pyrimidinediamine
Tylosin	1401-69-0	(11E,13E)-(4R,5S,6S,7R,9R,15R,16R)-15-[[(6-deoxy-2,3-di-O-methyl-β-D-allopyranosyl)oxy]methyl]-6-[[3,6-dideoxy-4-O-(2,6-dideoxy-3-C-methyl-α-L-ribo-hexopyranosyl)-3-(dimethylamino)-β-D-glucopyranosyl]oxy]-16-ethyl-4-hydroxy-5,9,13-trimethyl-7-(2-oxoethyl)oxacyclohexadeca-11,13-diene-2,10-dione

Acknowledgments

The authors would like to thank the UK Environment Agency and the European Commission who funded much of the work. The work for the European Commission was funded under the project ERAVMIS (project No. EVK-CT-1999-00003).

References

ADAS (1997) Animal manure practices in the pig industry: survey report. Prepared by ADAS Market Research Team, Wolverhampton England.

ADAS (1998) Animal manure practices in the dairy industry: survey report. Prepared by ADAS Market Research Team, Wolverhampton England.

Al-Ahmad A, Daschner FD, Kummerer K (1999) Biodegradability of ceftiofam, ciprofloxacin, meropenem, penicillin G and sulfamethoxazole and inhibition of waste water bacteria. Arch Environ Contam Toxicol 37:158–163.

Alderman DJ, Hastings TS (1998) Antibiotic use in aquaculture. Int J Food Sci Technol 33:139–155.

Arcand-Hoy LD, Nimrod AC, Benson WH (1998) Endocrine-modulating substances in the environment: estrogenic effects of pharmaceutical products. Int J Toxicol 17:139–158.

Armstrong A, Philips K (1998) A strategic review of sheep dipping. Environment Agency R & D Tech Rep P170. Environment Agency, Bristol, England.

Backhaus T, Grimme LH (1999) The toxicity of antibiotic agents to the luminescent bacterium *Vibrio fischeri*. Chemosphere 38:3291–3301.

Backhaus T, Scholze M, Grimme LH (2000) The single substance and mixture toxicity of quinolones to the bioluminescent bacterium *Vibrio fischeri*. Aquat Toxicol 49: 49–61.

Backhaus T, Faust M, Junghans M, Scholze M, Grimme H (2001) Low algal toxicities of quinolones confirm their specific molecular mechanism of action. Presented at the SETAC Europe 11th Annual Meeting, Madrid, May 2001.

Baguer AJ, Jensen J, Krogh PH (2000) Effects of the antibiotics oxytetracycline and tylosin on soil fauna. Chemosphere 40:751–758.

Bayer (1997) Baytril 10% injection: Safety Datasheet 345354/01. Bayer, Newbury, UK.

Björklund HV, Bylund G (1991) Comparative pharmacokinetics and bioavailability of oxolinic acid and oxytetracycline in rainbow trout (*Oncorhynchus mykiss*). Xenobiotica 21:1511–1520.

Björklund HV, Bondestam J, Bylund G (1990) Residues of oxytetracycline in wild fish and sediments from fish farms. Aquaculture 86:359–367.

Björklund HV, Råbergh CMI, Bylund G (1991) Residues of oxolinic acid and oxytetracycline in fish and sediments from fish farms. Aquaculture 97:85–96.

Bloom RA, Matheson JC (1993) Environmental assessment of avermectins by the US Food and Drug Administration. Vet Parasitol 48:281–294.

Bohm VR (1996) Auswirkungen von ruckstanden von antiinfektiva in tierischen ausscheidungen auf die gullebehandlung und den boden. Dtsch Tierearztl Wochenschr 103:237–284.

Bowen PDG (1995) Antiparasitic products. In: Animal Medicines: A User's Guide: Transport, Storage and Disposal of Animal Medicines. National Office of Animal Health Ltd, Enfield, UK.

Boxall ABA, Oakes D, Ripley P, Watts CD (2000) The application of predictive models in the environmental risk assessment of ECONOR. Chemosphere 40:775–782.

Boxall ABA, Blackwell P, Cavallo R, Kay P, Tolls J (2002) The sorption and transport of a sulfonamide antibiotic in soil systems. Toxicol Lett 131:19–28.

Briggs GG (1981) Theoretical and experimental relationships between soil adsorption, octanol-water partition coefficients, water solubilities, bioconcentration factors and the parachor. J Agric Food Chem 29:1050–1059.

Bull DL, Ivie GW, MacConnell JG, Gruber VF, Ku CC, Arison BH, Stevenson JM, VandenHeuve WJA (1984) Fate of avermectin B1a in soil and plants. J Agric Food Chem 32:94–102.

Burka JF, Hammell KL, Horsberg TE, Johnson GR, Rainnie DJ, Speare DJ (1997) Drugs used in salmonid aquaculture: a review. J Vet Pharmacol Ther 20:333–349.

Burkepile DE, Moore MT, Holland MM (2000) Susceptibility of five non-target organisms to aqueous diazinon exposure. Bull Environ Contam Toxicol 64:114–121.

Campbell WC, Fisher MH, Stapley EO, Albers-Schonberg G, Jacob TA (1983) Ivermectin: a potent new antiparasitic agent. Science 221:823–828.

Canavan A, Coyne R, Kennedy DG, Smith P (2000) Concentration of 22,23-dihydroavermectim B_{1a} detected in the sediments at an Atlantic salmon farm using orally administered ivermectin to control sea-lice infestation. Aquaculture 182:229–240.

Capone DG, Weston DP, Miller V, Shoemaker C (1996) Antibacterial residues in marine sediments and invertebrates following chemotherapy in aquaculture. Aquaculture 145:55–75.

Catherman DR, Szabo J, Batson DB, Cantor AH, Tucker RE, Mitchell JR (1991) Metabolism of narasin in chickens and quail. Poult Sci 70:120–125.

Chen Y, Rosazza JPN, Reese CP, Chang HY, Nowakowski MA, Kiplinger JP (1997) Microbial models of soil metabolim: biotransformations of danofloxacin. J Ind Microbiol Biotechnol 19:378–384.

Chien YH, Lai HT, Lui SM (1999) Modeling the effects of sodium chloride on degradation of chloramphenicol in aquaculture pond sediment. Sci Total Environ 239:81–87.

Chiu SHL, Green ML, Baylis FP, Eline D, Rosegay A, Meriwether H, Jacob TA (1990) Absorption, tissue distribution, and excretion of tritium-labelled ivermectin in cattle, sheep, and rat. J Agric Food Chem 38:2072–2076.

Committee for Veterinary Medicinal Products (CVMP) (1997) Note for guidance: environmental risk assessment for veterinary medicinal products other than GMO-containing and immunological products. EMEA/CVMP/055/96-FINAL. European Agency for the Evaluation of Medicinal Products, London.

Cook DF, Dadour IR, Ali DN (1996) Effect of diet on the excretion profile of ivermectin in cattle faeces. Int J Parasitol 26:291–295.

Cook RR (1995) Disposal of animal medicines. In: Animal Medicines: A User's Guide: Transport, Storage and Disposal of Animal Medicines. National Office of Animal Health Ltd, Enfield, UK.

Coyne R, Hiney M, O'Connor B, Kerry J, Cazabon D, Smith P (1994) Concentration and persistence of oxytetracycline in sediments under a marine salmon farm. Aquaculture 123:31–42.

Daughton CG, Ternes TA (1999) Pharmaceuticals and personal care products in the environment; agents of subtle change? Special report. Environ Health Perspect (Suppl) 107:907–938.

Davies IM, Rodger GK (2000) A review of the use of ivermectin as a treatment for sea lice (*Lepeophtheirus salmonis* (Kroyer) and *Caligus elongatus* Nordmann) infestation in farmed Atlantic salmon (*Salmo salar* L.). Aquacult Res 31:869–883.

Davies IM, McHenry JG, Rae GH (1997) Environmental risk of dissolved ivermectin to marine organisms. Aquaculture 158:63–275.

Davies IM, Gillibrand PA, McHenry JG, Rae GH (1998) Environmental risk of ivermectin to sediment-dwelling organisms. Aquaculture 163:29–46.

Davis ML, Lofthouse TJ, Stamm JM (1993) Aquatic photodegradation of 14C-sarafloxicin hydrochloride. Abstr Pap Am Chem S 205:91.

Donoho AL (1984) Biochemical studies on monensin. J Anim Sci 58:1528–1539.
Donoho AL (1987) Metabolism and residue studies with actaplanin. Drug Metab Rev 18:163–176.
Elanco (1998) Romensin G100 Premix: Material Safety Data Sheet. Elanco Animal Health, Speke, UK.
Elanco (2000) Apralan soluble powder: Material Safety Data Sheet: Elanco Animal Health, Speke, UK.
Environment Agency (EA) (1997) The occurrence of sheep-dip pesticides in environmental waters. National Centre for Toxic and Persistent Substances, Environment Agency, Peterborough.
Environment Agency (1998) Pesticides 1998: a summary of monitoring of the aquatic environment in England and Wales. National Centre for Ecotoxicology and Hazardous Substances, Environment Agency, Wallingford.
Environment Agency (1999) Sheep dip chemicals and textiles working group: a strategy for reducing sheep dip chemical pollution from the textile industry. National Centre for Ecotoxicology and Hazardous Substances, Environment Agency, Wallingford.
Environment Agency (2000) Welsh sheep dip monitoring programme 1999. Environment Agency Wales Cardiff. Environment Agency Midland Region, Solihull, April 2001.
Environment Agency (2001) Pesticides 1999/2000: a summary of monitoring of the aquatic environment in England and Wales. National Centre for Ecotoxicology and Hazardous Substances, Environment Agency, Wallingford.
Ervik A, Thorsen B, Eriksen V, Lunestad BT, Samuelsen OB (1994) Impact of administering antibacterial agents on wild fish and blue mussels *Mytilus edulis* in the vicinity of fish farms. Dis Aquat Org 18:45–51.
EU (1992) Council Directive 81/852/EEC of the Commission of September 28, 1981, as published in the Official Journal of the European Community of November 6, 1981, No. L317, page 16, amended by Directive 92/18/EEC of the Commission of March 20, 1992 as published in the Official Journal of the European Community of April 10, 1002, No. L97, p. 1.
FIDIN (2000) Veterinary therapeutic antibiotic use in the Netherlands: facts and figures. FIDIN presentation 12 April 2000, Societeit de Witte, the Hague.
Floate KD (1998) Off-target effects of ivermectin on insects and on dung degradationin southern Alberta, Canada. Bull Entomol Res 88:25–35.
Forbes AB (1993) A review of regional and temporal use of avemectins in cattle and horses worldwide. Vet Parasitol 48:19–28.
Gavalchin J, Katz SE (1994) The persistence of fecal-borne antibiotics in soil. J AOAC Int 77:481–485.
Gilbertson TJ, Hornish RE, Jaglan PS, Koshy T, Nappier JL, Stahl GL, Cazers AR, Nappier JM, Kubicek MF, Hoffman GA, Hamlow P (1990) Environmental fate of ceftiofur sodium, a cephalosporin antibiotic. Role of animal excreta in its decomposition. J Agric Food Chem 38:890–894.
Gover J, Strong L (1996) Determination of the toxicity of faeces of cattle treated with an ivermectin sustained-release bolus and preference trials using a dung fly, *Neomyia cornicina*. Entomol Exp Appl 81:133–139.
Grant A, Briggs AD (1998) Toxicity of ivermectin to estuarine and marine invertebrates. Mar Pollut Bull 36:540–541.
Grave K, Engelstad M, Søli NE, Toverud EL (1991) Clinical use of dichlorvos (Nuvan®) and trichlorfon (Neguvon®) in the treatment of salmon louse, *Lepeophtheirus*

salmonis. Compliance with the recommended treatment procedures. Acta Vet Scand 32:9–14.

Grave K, Lingaas E, Bangen M, Ronning M (1999) Surveillance of the overall consumption of antibacterial drugs in humans, domestic animals and farmed fish in Norway in 1992 and 1996. J Antimicrob Chemother 43:243–252.

Greiner P, Ronnefarth I (2001) Environmental risk assessment of veterinary medicines from a regulatory perspective. Paper presented at the SETAC Europe 11th Annual Meeting, Madrid, May 2001.

Gruber VF, Halley BA, Hwang S-C, Ku CC (1990) Mobility of avermectin B1a in soil. J Agric Food Chem 38:886–890.

Gustafson RH, Bowen RE (1997) Antibiotic use in animal agriculture. J Appl Microbiol 83:531–541.

Halley BA, Green ML, Chiu SHL (1986) Tissue depletion and metabolism of radiolabelled MK-0933 in cattle dosed percutaneously. Environmental assessment for the Ivomec © Pour-On Formulation, March 22, 1990 (unpublished).

Halley BA, Jacob TA, Lu AYH (1989) The environmental impact of the use of ivermectin: environmental effects and fate. Chemosphere 18:1543–1563.

Halley BA, VandenHeuvel WJA, Wislocki PG (1993) Environmental effects of the usage of avemectins in livestock. Vet Parasitol 48:109–125.

Halling-Sørenson B (1999) Algal toxicity of antibacterial agents used in intensive farming. Chemosphere 40:731–739.

Halling-Sørenson B, Nors Nielsen S, Lanzky PF, Ingerslev F, Holten Lützhøft HC, Jørgensen SE (1998) Occurrence, fate and effect of pharmaceutical substances in the environment: a review. Chemosphere 36:357–393.

Halling-Sørenson B, Jensen J, Tjørnelund J, Montforts MHMM (2001) Worst-case estimations of predicted environmental soil concentrations (PEC) of selected veterinary antibiotics and residues used in Danish agriculture. In: Kümmerer K (ed) Pharmaceuticals in the Environment: Sources, Fate, Effects and Risks. Springer, Heidelberg, pp 143–156.

Hamscher G, Abu-Quare A, Sczesny S, Höper H, Nau H (2000a) Determination of tetracyclines and tylosin in soil and water samples from agricultural areas in lower Saxony. In: van Ginkel LA, Ruiter A (eds) Proceedings of the Euroresidue IV Conference, Veldhoven, Netherlands, 8–10 May 2000. National Institute of Public Health and the Environment (RIVM), Bilthoven, Netherlands, pp 522–526.

Hamscher G, Sczesny S, Abu-Quare A, Höper H, Nau H (2000b) Substances with pharmacological effects including hormonally active substances in the environment: identification of tetracyclines in soil fertilised with animal slurry. Dtsch Tierärztl Wochenschr 107:293–348.

Hamscher G, Sczesny S, Höper H, Nau H (2000c) Tetracycline and chlortetracycline residues in soil fertilized with liquid manure. In: Hartung J, Wathes C (2000) (eds) Livestock Farming and the Environment, Sonderheft 226. Braunschweig, Germany, pp 27–31.

Hansen PK, Lunestad BT, Samuelsen OB (1993) Effects of oxytetracycline, oxolinic acid and flumequine on bacteria in an artificial marine fish farm sediment. Can J Microbiol 39:1307–1312.

Harries JE, Sheahan DA, Jobling S, Matthiessen P, Neall P, Sumpter JP, Taylor T, Zaman N (1997) Estrogenic activity in five United Kingdom rivers detected by the measurement of vitellogenesis in caged male trout. Environ Toxicol Chem 16:534–542.

Hektoen H, Berge JA, Hormazabal V, Yndestad M (1995) Persistence of antibacterial agents in marine sediments. Aquaculture 133:175–184.

Hirsch R, Ternes T, Haberer K, Kratz KL (1999) Occurrence of antibiotics in the aquatic environment. Sci Total Environ 225:109–118.

Holm JV, Rügge K, Bjerg PL, Christensen TH (1995) Occurrence and distribution of pharmaceutical organic compounds in the groundwater downgradient of a landfill (Grinsted, Denmark). Environ Sci Technol 29:1415–1420.

Holten Lützhøft HC, Halling-Sørensen B, Jørgensen SE (1999) Algal toxicity of antibacterial agents applied in Danish fish farming. Arch Environ Contam Toxicol 36:1–6.

Høy T, Horsberg TE, Nafstad I (1990) The disposition of ivermectin in Atlantic salmon (*Salmo salar*). Pharmacol Toxicol 67:307–312.

HSE (1997) Sheep dipping. Booklet AS29 (revision 2). Health and Safety Executive, London.

Hustvedt SO, Salte R, Kvendseth O, Vassvik V (1991) Bioavailability of oxolinic acid in Atlantic salmon (*Salmo salar* L.) from medicated feed. Aquaculture 97:305–310.

Ingerslev F, Halling-Sørensen B (2001) Biodegradability of metronidazole, olaquindox and tylosin and formation of tylosin degradation products in aerobic soil/manure slurries. Chemosphere 48:311–320.

Intervet (1999) Salocin: Safety Data Sheet. Intervet, Milton Keynes, UK.

Jacobsen P, Berglind L (1988) Persistence of oxytetracycline in sediments from fish farms. Aquaculture 70:365–370.

Jernigan AP, Herd RP, Sams R (1990) Determination of ivermectin in equine feces. Fed Am Soc Exp Biol 4:Abstract 2810.

Jørgensen SE, Halling-Sørensen B (2000) Drugs in the environment. Chemosphere 40:691–699.

Kerry J, Coyne R, Gilroy D, Hiney M, Smith P (1996) Spatial distribution of oxytetracycline and elevated frequencies of oxytetracycline resistance in sediments beneath a marine salmon farm following oxytetracycline therapy. Aquaculture 145:31–39.

Kolpin DW, Furlong ET, Meyer MT, Thurman EM, Zaugg SD, Barber LB, Buxton HT (2002) Pharmaceuticals, hormones, and other organic wastewater contaminants in US streams 1999–2000: a national reconnaissance. Environ Sci Technol 36:1202–1211.

Kopf W (1995) Wirkung endokriner stoffe in biotests mit wasserorganismen. Vortrag bei der 50. Fachtung des bay, LA fu wasserwirtschaft: stoffe mit endokriner. Wirkung im Wasser (abstract).

Kummerer K, Al-Ahmed A, Mersch-Sundermann V (2000) Biodegradation of some antibiotics, elimination of the genotoxicity and affection of wastewater bacteria in a simple test. Chemosphere 40:701–710.

Lai HT, Liu SM, Chien YH (1995) Transformation of chloramphenicol and oxytetracycline in aquaculture pond sediments. J Environ Sci Health A 30:1897–1923.

Lanzky PF, Halling-Sørensen B (1997) The toxic effect of the antibiotic metronidazole on aquatic organisms. Chemosphere 35:2553–2561.

Larkin DJ, Tjeerdema RS (2000) Fate and effects of diazinon. Rev Environ Contam Toxicol 166:49–82.

Lewis KA, Bardon KS (1998) A computer-based informal environmental management system for agriculture. Environ Model Software 13:123–137.

Lewis S (1998) Proposed environmental quality standards for sheep dip chemicals in water. Chlorfenvinphos, Coumaphos, Diazinon, Fenchlorphos, Flumethrin and Propetamphos—an update. Draft R & D Report P128. Environment Agency, Bristol, England.

Lewis S, Watson A, Hedgecott S (1993) Propsed environmental quality standards for sheep dip chemicals in water. Chlorfenvinphos, Coumaphos, Diazinon, Fenchlorphos, Flumethrin and Propetamphos. WRc plc, R & D Note 216. Scotland and Northern Ireland Forum for Environmental Research and the National Rivers Authority, Bristol, England.

Liddel JS (2001) Sheep ectoparasiticide use in the UK: 1993, 1997 and 1999. Paper presented to the 5th International Sheep Veterinary Congress, Stellenbosch, South Africa. 21–25 January

Littlejohn JW, Melvin AAL (1991) Sheep-dips as a source of pollution of freshwaters: a study in Grampian Region. Journal of the Chartered Institute of Water and Environmental Management 5:21–27.

Loke ML, Ingerslev F, Halling-Sørensen B, Tjornelund J (2000) Stability of tylosin A in manure containing test systems determined by high performance liquid chromatography. Chemosphere 40:759–765.

Lunestad BT (1992) Fate and effects of antibacterial agents in aquatic environments. In: Proceedings of the Conference on Chemotherapy in Aquaculture: From Theory to Reality. Office International des Epizooties, Paris, France. p. 152–161.

Lunestad BT, Samuelsen OB, Fjelde S, Ervik A (1995) Photostability of eight antibacterial agents in seawater. Aquaculture 134:217–225.

Madsen M, Overgaard Nielsen B, Holter P, Pedersen OC, Brrochner Jespersen J, Vagn Jensen K-M, Nansen P, Gronvold J (1990) Treating cattle with ivermectins: effects on fauna and decomposition of dung pats. J Appl Ecol 27:1–15.

MAFF (1998) Code of good agricultural practice for the protection of water. Ministry of Agriculture Fisheries and Food, Welsh Office Agricultural Department, Cardiff, Wales.

Magnussen JD, Dalidowicz JE, Thomson TD, Donoho AL (1991) Tissue residues and metabolism of avilamycin in swine and rats. J Agric Food Chem 39:306–310.

Marengo JR, Kok RA, O'Brien K, Velagaleti R, Stamm JM (1997) Aerobic biodegradation of (^{14}C)-sarafloxicin hydrochloride in soil. Environ Toxicol Chem 16:462–471.

Matha V, Weiser J (1988) The molluscicidal effect of ivermectin on *Biomphalaria glabrafa*. J Inveretebr Pathol 52:354–355.

McCracken DI (1993) The potential for avermectins to affect wildlife. Vet Parasitol 48: 273–280.

McKellar QA (1997) Ecotoxicology and residues of anthelmintic compounds. Vet Parasitol, 72:413–435.

Melancon SM, Pollard JE, Hern SC (1986) Evaluation of SESOIL, PRZM and PESTAN in a laboratory column leaching experiment. Environ Toxicol Chem 5:865–878.

Merck, Sharp & Dohme (1983) Environmental impact report on Ivomec injection. New Anim Drug Application No.128–409. Rahway, NJ, USA.

Merial (1998) Eprinex pour-on for beef and dairy cattle: Product Safety Information Sheet. Merial Animal Health Ltd, Harlow, UK.

Meyer MT, Bumgarner JE, Varns JL, Daughtridge JV, Thurman EM, Hostetler KA (2000) Use of radioimmunoassay as a screen for antibiotics in confined animal feeding operations and confirmation by liquid chromatography/mass spectrometry. Sci Total Environ 248:181–187.

Meyer W, Backhaus T, Froehner K, Scholze M, Grimme H (2001) Single substance and mixture toxicity of selected pharmaceuticals. Presented at the SETAC Europe 11th Annual Meeting, Madrid, May 2001.

Migliore L, Brambilla G, Cozzolino S, Gaudio L (1995) Effect on plants of sulphadimethoxine used in intensive farming (*Panicum miliaceum, Pisum sativum* and *Zea mays*). Agric Ecosyst Environ 52:103–110.

Migliore L, Brambilla G, Casoria P, Civitareale SC, Gaudio L (1996) Effect of sulphadimethoxine contamination on barley (*Hordeum disticum* L., Poaceae, Liliopsida). Agric Ecosyst Environ 60:121–128.

Migliore L, Civitareale C, Brambilla G, Dojmi Di Delupis G (1997a) Toxicity of several important agricultural antibiotics to Artemia. Water Res 31:1801–1806.

Migliore L, Civitareale C, Brambilla SC, Casoria P, Gaudio L (1997b) Effects of sulphadimethoxine on cosmopolitan weeds (*Ameranthus retroflexus* L., *Plantago major* L. and *Rumex acetosella* L.). Agric Ecosyst Environ 65:163–168.

Migliore L, Cozzolino S, Fiori M (2000) Phytotoxicity to and uptake of flumequine used in intensive aquaculture on the aquatic weed, *Lythrum salicaria* L. Chemosphere 40:741–750.

Miller JA, Kunz SE, Oehlar DD, Miller RW (1981) Larvicidal activity of Merck MK-933, an avermectin, against the horn fly, stable fly, face fly, and house fly. J Econ Entomol 74:608–611.

Montforts MHMM (1999) Environmental risk assessment for veterinary medicinal products. Part 1: other than GMO-containing and immunological products. RIVM report 601300 001. National Institute of Public Health and the Environment, Bilthoven, The Netherlands.

Nessel RJ, Wallace DH, Wehner TA, Tait WE, Gomez L (1989) Environmental fate of ivermectin in a cattle feedlot. Chemosphere 18:1531–1541.

Novartis (1999) Safety Data Sheet: Sulfachloropyridazine-Na. Release date 18 Oct 1999. Basel, Switzerland.

Nowara A, Burhenne J, Spiteller M (1997) Binding of fluoroquinolone carboxylic acid derivatives to clay minerals. J Agric Food Chem 45:1459–1463.

Oka H, Ikai Y, Kawamura N, Yamada M, Harada K-I, Ito S, Suzuki, M (1989) Photodecomposition products of tetracycline in aqueous solution. J Agric Food Chem 37:226–231.

Pelicaan CHP, van Turnhout J, Pijpers A (2000) Verbruikscijfers van antibacteriële middelen bij landbouwhuisdieren. Poster obtained from University of Utrecht, The Netherlands.

Pepper T, Carter A (2000) Monitoring of pesticides in the environment. Report prepared for the Pesticides in the Environment Working Group. R & D Project E1-076. Bristol, England.

Pouliquen H, Le Bris H, Pinault L (1992) Experimental study of the therapeutic application of oxytetracycline, its attenuation in sediment and sea water, and implication for farm culture of benthic organisms. Mar Ecol Prog Ser 89:93–98.

Purdom CE, Hardiman PA, Bye VA, Eno NC, Tyler CR, Sumpter JP (1994) Oestrogenic effects of effluents from sewage treatment works. Chem Ecol 8:275–285.

Rabølle M, Spliid NH (2000) Sorption and mobility of metronidazole, olaquindox, oxytetracycline and tylosin in soil. Chemosphere 40:715–722.

Richardson MI, Bowron JM (1985) The fate of pharmaceutical chemicals in the aquatic environment. J Pharm Pharmacol 37:1–12.

Ridsdill-Smith TJ (1988) Survival and reproduction of *Musca vetustissima* Walker (Diptera: Muscidae) and a scarabaeine dung beetle in dung cattle treated with avermectin B1. J Aust Entomol Soc 27:175–178.

Ridsdill-Smith TJ (1993) Effects of avermectin residues in cattle dung on dung beetle (Coleoptera: Scarabaeidae) reproduction and survival. Vet Parasitol 48:127–137.

Samuelsen OB (1989) Degradation of oxytetracycline in seawater at two different temperatures and light intensities and the persistence of oxytetracycline in the sediment from a fish farm. Aquaculture 83:7–16

Samuelsen OB, Solheim E, Lunestad BT (1991) Fate and microbiological effects of furazolidone in a marine aquaculture sediment. Sci Total Environ 108:275–283.

Samuelsen OB, Lunestad BT, Husevåg B, Hølleland T, Ervik A (1992a) Residues of oxolinic acid in wild fauna following medication in fish farms. Dis Aqua Org 12: 111–119.

Samuelsen OB, Torsvik V, Ervik A (1992b) Long-range changes in oxytetracycline concentration and bacterial resistance towards oxytetracycline in a fish farm sediment after medication. Sci Total Environ 114:25–36.

Samuelsen OB, Lunestad BT, Ervik A, Fjelde S (1994) Stability of antibacterial agents in an artificial marine aquaculture sediment studied under laboratory conditions. Aquaculture 126:283–290.

Schmidt CD (1983) Activity of an avermectin against selected insects in aging manure. Environ Entomol 12:455–457.

Schweinfurth H, Lange R, Gunzel P (1996) Environmental fate and ecological effects of steroidal estrogens. IBC Conference Proceedings "Oestrogenic Compounds in the Environment," London, 9–10 May 1996.

SEPA (1999) Enamectin benzoate an environmental risk assessment. SEPA 66/99. Report of the SEPA Fish Farm Advisory Group, East Kilbride, Scotland.

SEPA (2000) Long-term biological monitoring trends in the Tay System 1988–1999. Scottish Environment Protection Agency, Eastern Region, Scotland.

Shore LS, Shemesh M, Cohen R (1988) The role of oestradial and oestrone in chicken manure silage in hyperoestrogenism in cattle. Aust Vet J 65:67.

Smith P (1996) Is sediment deposition the dominant fate of oxytetracycline used in marine salmonid farms: a review of available evidence. Aquaculture 146:157–169.

Sommer C, Overgaard Nielsen B (1992) Larvae of the dung beetle *Onthophagus gazella* F. (Col., Scarabaeidae) exposed to lethal and sublethal ivermectin concentrations. J Appl Entomol 114:502–509.

Sommer C, Steffansen B (1993) Changes with time after treatment in the concentrations of ivermectin in fresh cow dung and in cow pats aged in the field. Vet Parasitol 48: 67–73.

Sommer C, Steffansen B, Overgaard Nielsen B, Grønvold J, Kagn-Jensen KM, Brøchner Jespersen J, Springborg J, Nansen P (1992) Ivermectin excreted in cattle dung after subcutaneous injection or pour-on treatment:concentrations and impact on dung fauna. Bull Entomol Res 82:257–264.

Sommer C, Gronvold J, Holter P, Nansen P (1993) Effect of ivermectin on two afrotropical beetles *Onthophagus gazella* and *Diastellopalpus quinquedens* (Coleoptera: Scarabaeidae). Vet Parasitol 48:171–179.

Spaepen KRI, Leemput LJJ, Wislocki PG, Verschueren C (1997) A uniform procedure to estimate the predicted environmental concentration of the residues of veterinary medicines in soil. Environ Toxicol Chem 16:1977–1982.

Stout SJ, Wu J, daCunha AR, King KG, Lee A (1991) Maduramycin α: characterisation of ^{14}C-derived residues in turkey excreta. J Agric Food Chem 39:386–391.

Strong L (1992) Avermectins: a review of their impact on insects of cattle dung. Bull Entomol Res 82:265–274.

Strong L (1993) Overview: the impact of avermectins on pastureland ecology. Vet Parasitol 48:3–17.

Strong L, Brown TA (1987) Avermectins in insect control and biology: a review. Bull Entomol Res 77:357–389.

Strong L, James S (1993) Some effects of ivermectin on the yellow dung fly, *Scatophaga stercoraria*. Vet Parasitol 48:181–191.

Strong L, Wall R (1994) Effects of ivermectin and moxidectin on the insects of cattle dung. Bull Entomol Res 84:403–409.

Taylor SM (1999) Sheep scab—environmental considerations of treatment with doramectin. Vet Parasitol 83:309–317.

Ternes TA, Stumpf M, Mueller J, Haberer K, Wilken R-D, Servos M (1999) Behaviour and occurrence of estrogens in municipal sewage treatment plants. I. Investigations in Germany, Canada and Brazil. Sci Total Environ 225:81–90.

Thain JE, Davies IM, Rae GH, Allen YT (1997) Acute toxicity of ivermectin to the lugworm *Arenicola marina*. Aquaculture 159:47–52.

Thorpe JE, Talbot C, Miles MS, Rawlings C, Keay DS (1990) Food consumption in 24 hours by Atlantic salmon (*Salmo salar* L) in a sea-cage. Aquaculture 90:41–47.

Thurman EM, Lindsey ME (2000) Transport of antibiotics in soil and their potential for groundwater contamination. Poster presented at The SETAC World Congress, Brighton, May 2000.

Tomlin CDS (1997) The Pesticide Manual, 11th Ed. BCPC, Farnham, Surrey, UK.

Turner MJ, Schaeffer JM (1989) Mode of action of ivermectin. In: Campbell WC (ed) Ivermectin and Abamectin. Springer-Verlag, New York, pp 73–88.

USEPA (1997) Profile of the pharmaceutical manufacturing industry. EPA/310-R-97-005. U.S. Environmental Protection Agency, Office of Compliance, Washington, DC.

USEPA (2001) ECOTOX Database System. Prepared for USEPA, Office of Research and Development, National Health and Environmental Effects Research Laboratory, Mid-Continent Ecology Division (MED), Duluth, MN, by OAO Corporation, Duluth, MN.

van Dijk J, Keukens HJ (2000) The stability of some veterinary drugs and coccidiostats during composting and storage of laying hen and broiler faeces. In: van Ginkel LA, Ruiter A (eds) Residues of Veterinary Drugs in Food. Proceedings of the Euroresidue IV Conference, Veldhoven, The Netherlands, 8–10 May, 2000.

Velagaleti R, Gill M (2003) Degradation and depletion of pharmaceuticals in the environment. In: Daughton CG, Ternes TA (eds) Proceedings of the American Chemical Society: Issues in the Analysis of Environmental Endocrine Disruptors, San Francisco, CA, 27 March, 2000.

Velagaleti RR, Davis ML, O'Brien GK (1993) The bioavailability of ^{14}C-sarafloxicin hydrochloride in three soils and a marine sediment as determined by biodegradation and sorption/desorption parameters. Poster presented at the American Chemical Society E-Fate Meeting, 28 March, 1993, Abstr Pap Am Chem S 205:92.

VICH (unpublished) Analysis of data and information to support a PEC soil trigger value for Phase I.

VICH (2000) Environmental impact assessment (EIAs) for veterinary medicinal products (VMPs): Phase 1. VICH GL6 (Ecotoxicity Phase 1), June 2000, For Implementation at Phase 7. Brussels, Belgium.

Vink K, Dewi L, Bedaux J, Tompot A, Hermans M, VanStraalen NM (1995) The importance of the exposure route when testing the toxicity of pesticides to saprotrophic isopods. Environ Toxicol Chem 14:1225–1232.

Virtue WA, Clayton JW (1997) Sheep Dip Chemicals and Water Pollution Sci Tot Env 194/195:207–217.

VMD (2001) Sales of antimicrobial products used as veterinary medicines and growth promoters in the UK in 1999. Veterinary Medicines Directorate, Addlestone, UK.

Wall R, Strong L (1987) Environmental consequences of treating cattle with the antiparasitic drug ivermectin. Nature (Lond) 327:418–421.

Warman PR (1980) The effect of amprolium and aureomycin on the nitrification of poultry manure-amended soil. J Soil Sci Soc Am 44:1333–1334.

Warman PR, Thomas RL (1981) Chlortetracycline in soil amended with poultry manure. Can J Soil Sci 61:161–163.

Webb SF (2001) A data-based perspective on the environmental risk assessment of human pharmaceuticals. I. Collation of available ecotoxicity data. In: Kümmerer K (ed) Pharmaceuticals in the Environment: Sources, Fate, Effects and Risks. Springer, Heidelberg. pp. 175–201.

Weereasinghe CA, Towner D (1997) Aerobic biodegradation of virginamycin in soil. Environ Toxicol Chem 16:1873–1876.

Wollenberger L, Halling-Sørensen B, Kusk KO (2000) Acute and chronic toxicity of veterinary antibiotics to *Daphnia magna*. Chemosphere 40:723–730.

WRc-NSF (2000) The development of a model for estimating the environmental concentrations (PECs) of veterinary medicines in soil following manure spreading (Project Code VM0295). Final Project Report to MAFF, London, England.

Yeager RL, Halley BA (1990) Sorption/desorption of [^{14}C] efrotomycin with soils. J Agric Food Chem 38:886–890.

Manuscript received November 18; accepted Dec 23, 2002.

Health Effects of *Acanthamoeba* spp. and Its Potential for Waterborne Transmission

Nena Nwachuku and Charles P. Gerba

Contents

I. Introduction	94
II. General Characteristics	96
A. Methods of Identification	96
B. Cultivation	97
C. Significance of Endosymbiosis	98
III. Occurrence	99
A. Surface Waters	100
B. Tapwater and Bottled Water	101
C. Swimming Pools and Spas	102
D. Sewage and Biosolids	102
E. Animal Wastes	102
F. Air, Dust, and Soil	103
IV. Health Effects	103
A. Eye Infections (Acanthamoebic Keratitis	104
B. Granulomatous Amoebic Encephalitis	111
C. Other Infections caused by *Acanthamoeba*	114
D. Immunocompromised Individuals	114
E. Effect of Endosymbiosis on Virulence	114
V. Risk Assessment	115
A. Resistance to Drinking Water Treatment and Disinfection	117
B. Dose Response	117
C. Risk Characterization	118
VI. Association of Contact Lenses with Acanthamoebic Keratitis	119
A. Types of Contact Lenses	119
B. Demographics of Contact Lens Use	119
C. Risk Factors	121
VII. Summary	122
Acknowledgments	122
References	123

Communicated by George W. Ware.

N. Nwachuku (✉)
Office of Science and Technology, Office of Water, U.S. Environmental Protection Agency, 1200 Pennsylvania Ave. N.W., Mc 4304T, Washington, DC 20460, U.S.A.

C.P. Gerba
Department of Soil, Water and Environmental Science, University of Arizona, Tucson, AZ 85721, U.S.A.

I. Introduction

The Safe Drinking Water Act, as amended in 1996, requires the U.S. Environmental Protection Agency (EPA) to publish a Drinking Water Contaminant Candidate List (CCL). During the development of the first draft list in 1996, EPA obtained input from stakeholders including an international panel of expert microbiologists and the Science Advisory Board. The expert microbiologists panel recommended that EPA issue a public health guidance for controlling *Acanthamoeba* eye infection in contact lens wearers. *Acanthamoeba* spp. are protozoan common in water and soil that have been associated with inflammation of the human cornea, usually in contact lens wearers, and with chronic encephalitis in immuno deficient individuals. The organism is transmitted by contact of the eye (or possibly other body surfaces) with contaminated water, air, or soil. There is no evidence that it is transmitted by ingestion. The goal of this review is to assess the health effects of *Acanthamoeba* and the significance of water in its transmission.

Acanthamoeba is a protozoan genus. Protozoa are unicellular eukaryotic animals and, although they are widespread in the environment, only a few are capable of causing disease in humans. Several of the pathogenic protozoa are transmitted by water, including *Giardia lamblia*, *Cryptosporidium* spp., *Naegleria fowleri*, and certain *Acanthamoeba* spp (Table 1). *Acanthamoeba* are free-living amoebae that have no defined shape. They move by pseudopods, extensions of the cell membrane into which the cytoplasm moves. They normally live in soil, freshwater, brackish water, sewage, and biosolids, feeding on bacteria and multiplying in their environmental niche as free-living organisms. They are capable of causing infections of the human skin, lungs, eye, and brain and can feed on human tissue. Because of their ability to live both free in nature and as pathogens in a host, they are also called amphizoic amoebe; this is in contrast to *Giardia* and *Cryptosporidium*, which do not replicate in the environment (Table 1). These waterborne pathogenic protozoa are transmitted by ingestion and replicate only within the host.

The genus *Acanthamoeba* consists of as many as 20 species classified in three groups based on cyst morphology (Table 2). Unlike *Naegleria fowleri*, the

Table 1. Waterborne/water-based pathogenic protozoa.

Type	Genus/species	Disease/symptoms
Amoeboid	*Acanthamoeba*	Eye infection (keratitis), brain infection (meningoencephalitis)
	Naegleria	Brain infection (meningoencephalitis)
	Entamoeba histolytica	Amoebic diarrhea (liver abcess)
Flagellate	*Giardia lamblia*	Diarrhea
Apicomplexan	*Toxoplasma gondii*	Fever, loss of fetus
	Cryptosporidium	Diarrhea
	Cyclospora cayetanesis	Diarrhea

Table 2. Currently identified species of *Acanthamoeba*.

A. astronyxis
A. castellanii
A. comandoni
A. culbertsoni
A. divionensis
A. echinulata
A. gigantea
A. griffini
A. hatchetti
A. healyi
A. jacobsi
A. lenticulata
A. lugdunensis
A. mauritaniensis
A. palestinensis
A. paradivionensis
A. pearcei
A. polyphaga
A. quina
A. rhisodes
A. royreba
A. stevensoni
A. terricola
A. triangularis
A. tubiashi

most important species of *Naegleria* that causes human disease, several species of *Acanthamoeba* are known to cause infections in humans. These species include *A. astronyxis, A. castellanii, A. culbertsoni, A. divionensis, A. healyi, A. rhisodes, A. hatchetti, A. palestinensis*, and *A. polyphaga*. Exposure to contaminated recreational water and tapwater has been implicated as a source of exposure, especially for those species causing infections of the eye.

The taxonomy of *Acanthamoeba* is a contentious area. Those species now known as *Acanthamoeba* were previously placed in the genus *Hartmanella*, but in 1967 they were definitely classified as a separate genus by Page (1967). Pussard and Pons (1977) later proposed a classification based mainly on cyst morphology that identified 18 species (Table 2). The species were classified into three morphological groups (Table 3).

Group I has large cysts with rounded outer walls (ectocysts) that are clearly separated from the inner walls (endocysts). The inner and outer walls are joined, forming a star-shaped structure. Group II cysts are smaller, with variable endocyst shapes. Group III cysts are smaller than Group II cysts, with poorly separated walls. The major human pathogens belong to Group II, although *A. culbertsoni*, from Group III, is also a recognized pathogen.

Table 3. *Acanthamoeba* species classification.

Group I	Group II	Group III
A. astronyxis	*A. castellani*	*A. palastinensis*
A. comandoni	*A. mauritaniensis*	*A. culbertsoni*
A. echinulata	*A. polyphaga*	*A. lenticulata*
	A. lugdunesis	*A. pustulosa*
	A. quina	*A. royreba*
	A. rhisodes	
	A. divionensis	
	A. paradivionensis	
	A. griffini	
	A. triangularis	

Source: Pussard and Pons (1977).

II. General Characteristics

Acanthamoeba has two stages in its life cycle: the trophozoite and the cyst (Fig. 1). *Acanthamoeba* trophozoites measure 15–45 µm and are characterized by the presence of fine, tapering, spinelike projections from the surface of the body called acanthopodia. The acanthopodia can be periodically protruded and retracted. The trophozoites usually have one nucleus with a large, dense nucleolus. *Acanthamoeba* divide by conventional mitosis, in which the nucleolus and the nuclear membrane disappear during cell division. The trophozoite feeds on bacteria by engulfing them (phagocytosis).

Under adverse environmental conditions, a dormant cyst is formed that is resistant to desiccation, temperature extremes, and disinfectants. The cyst may remain viable for many years and, when it is exposed to a food source, it again assumes the trophozoite form. It is not understood how the cyst recognizes a food source. It will readily excyst in the presence of both liquid nutrients and bacteria.

Acanthamoeba are carriers of intracellular bacteria, especially *Legionella* species, which have the ability to reproduce within the trophozoite. It has been proposed that this may be of importance in the persistence and spread of these organisms in the environment (King et al. 1988).

A. Methods of Identification

The identification of individual *Acanthamoeba* is based on morphological observations, but taxonomic studies have employed isoenzyme (de Jonckheere 1987) or mitochondrial species DNA restriction endonuclease analysis in an attempt to form a classification system. A study of mitochondrial DNA has produced comparable results. In the first study, 33 strains, of which 30 were corneal isolates, were separated into 10 groups according to restriction length pattern polymorphism.

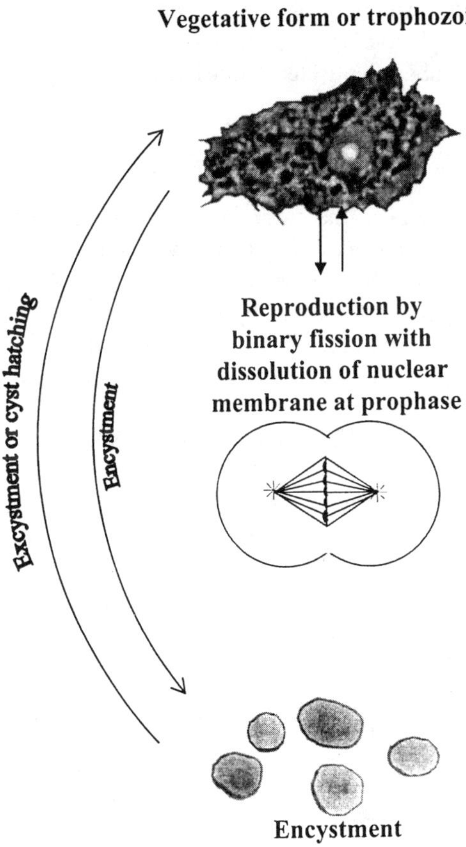

Vegetative form or trophozoite

Reproduction by binary fission with dissolution of nuclear membrane at prophase

Excystment or cyst hatching

Encystment

Locomotive or proliferative form

Encystment

Fig. 1. Life cycle of *Acanthamoeba*.

B. Cultivation

Acanthamoeba are easily grown on non-nutrient agar plates seeded with *Escherichia coli* or *Klebsiella pneumoniae* (Kilvington et al. 1990; Visvesara et al. 1975). One of the more common methods is to smear or streak a suitable bacterial food organism such as *Escherichia coli* or *Klebsiella pneumoniae* over the agar surface, seal the plates with tape, invert them, and incubate them in boxes lined with wet paper towels to maintain humidity. *Acanthamoeba* will migrate across the plate using bacteria as a food source. Overproliferation of bacteria is prevented by the nonnutrient agar. With incubation at 32 °C, the migration tracks of the amoebae are usually easily visible within 48 hr, but occasionally longer incubation (up to 2 wk) is needed (Illingworth and Cook 1998).

Formulations for several complex liquid axenic (bacteria-free) media may be found in a publication by the American Type Culture Collection (Nerad 1993). Because some species of amphizoic amoebe grow at mammalian body temperatures, many laboratories incubate replicate cultures at room temperature, 37°–45 °C, or higher.

C. Significance of Endosymbiosis

Acanthamoeba feeds on bacteria in the environment, trapping them within its cytoplasm, a process known as phagocytosis. Phagocytosed bacteria are usually killed and digested by the amoebae; however, some species of bacteria may grow and reproduce within the cytoplasm and become symbionts. Symbiotic relationships are beneficial to both organisms. When the bacteria have adapted to the intercellular environment of the protozoan host, the event is referred to as endosymbiosis. Both the survival and virulence of both organisms may be enhanced by this relationship (see IV.E). Rowbotham (1980) first reported the association of the amoebae *Naegleria* and *Acanthamoeba* with the symbiont *Legionella pneumophila*, the causative agent of legionnaires' disease. Several species of free-living amoebae have been shown to support the growth of legionellas (Fields 1993), and environmental growth of legionellas in the absence of protozoa has not been documented. It is thought that the protozoa are the primary means of proliferation of these bacteria under natural conditions (Fields et al. 1989; Hay et al. 1995). This endosymbiotic relationship can modify the virulence of *Legionella* (Dowling et al. 1992). It may also be involved in the observed phenomenon that *L. pneumophila* can be viable but nondetectable by cultivation on agar-based systems (Connor et al. 1993). Hay and Seal (1994b) have proposed that the latter observation may have profound implications with regard to surveillance of water systems for *Legionella*, especially with prevention of outbreaks of nosocomial legionnaires' disease.

Various waterborne pathogens have been shown to develop an endosymbiotic relationship. The spectrum of pathogens able to survive and multiply to various degrees within *Acanthamoeba* is given in Table 4. For all the organisms, *Acan-*

Table 4. Bacterial endosymbionts[a] of *Acanthamoeba*.

Legionella pneumophila
Mycobacterium avium
Burkholderia picketti
Vibrio cholerae
Francisella tularensis
Chlamydia pneumoniae
Rickettsiales
Listeria monocytogenes

[a]Live within the *Acanthamoeba*.
Source: Fritsche et al. 1999; Ly and Muller, 1990.

thamoeba are potential reservoirs and vectors, in part because of their ubiquity in the environment, their resistant cyst stages, and their potential to grow in water supplies, cooling, humidification systems, and recreational waters.

Endosymbiosis has also been shown to protect *Legionella* against disinfection (Kilvington and Price 1990) and to enhance the ability of both the bacteria and protozoa to cause disease (see IV.E). Thus, the presence of *Acanthamoeba* in drinking water distribution systems not only may add to the survival of other waterborne pathogens, but this relationship may enhance their virulence (Fig. 2).

III. Occurrence

Acanthamoeba are abundant in the environment and have been isolated from tapwater, seawater, air, soil, dust, and vegetables (Table 5). They feed on bacteria, fungi, other protozoa, and Cyanobacteria (blue-green algae) (Rodriguez-Zaragoza 1994). They are found in greatest numbers where other microorganisms are most numerous.

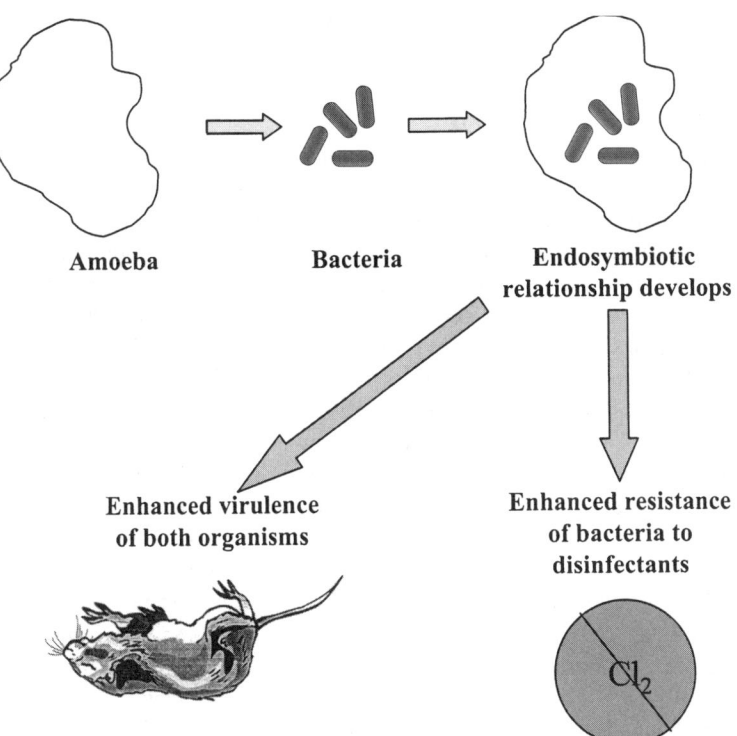

Fig. 2. Significance of endosymbiosis to waterborne disease transmission.

Table 5. Occurrence of *Acanthamoeba*.

Source	Reference
Water fountains	Crespo et al. (1990)
Tap water (Mexico)	Rivera et al. (1979)
Bottled water (Mexico)	Rivera et al. (1981)
Hospital tap water	Rohr et al. (1998)
Eyewash stations	Tyndall et al. (1987)
Freshwater ponds	John and Howard (1995)
Thermal water	de Jonckheere (1979a); Dive et al. (1982)
Well water	Jones et al. (1975)
Physiotherapy tubs	Penas-Ares et al. (1994)
Aquaria	de Jonckheere (1979b)
Municipal sewage	Singh and Das (1972)
Ocean sewage dump site	Sawyer et al. (1982)
House dust	Yamaura et al. (1993)
Garden soil	Singh (1952)
Sand box	Yamaura et al. (1993)
Garden vegetables	Rude et al. (1984)
Fish	Taylor (1977)
Air conditioner	Walker et al. (1986)

A. Surface Waters

One of the early studies on the numbers of *Acanthamoeba* in a freshwater lake, published by O'Dell (1979), reported a distinct seasonal variation in populations of *A. polyphaga* ranging from approximately 200 to 1000 per gram of lake-bottom mud during February through July and 200–2100/g during August through January. Peak counts were noted during August and September. *Acanthamoeba castellanii* was also observed in this study but was recovered only on three occasions and did not exceed a population of 200/g. Detterline and Wilhelm (1991) collected water samples from 59 sites in federally managed recreational waters of the United States and recovered temperature-tolerant strains of *Acanthamoeba* from 16 of 31 sites that grew at 37 °C. Kyle and Noblet (1987) published a detailed account of amoebae present in a spillway reservoir in South Carolina. The authors studied the lake throughout the course of a year to record seasonal influences on amoeba populations, such as dissolved oxygen, attenuation, and water temperature. In the surface water, amphizoic amoebae ranged from 5 to 10 amoebae/50 mL water in May and peaked at 98/50 mL in July.

Asiri et al. (1990) tested sediments along a transect in the Potomac River ranging from nontidal waters above Washington, DC to tidal waters (brackish) 0.8 m below a municipal sewage treatment plant. They identified seven species of *Acanthamoeba*, most of which occurred in the tidal portion of the river near the sewage treatment plant. John and Howard (1995) processed 2016 samples from ponds in Oklahoma and recovered 34 strains of pathogenic (induced brain damage) amoebae with 35% identified as *Acanthamoeba*. They estimated approximately 1 pathogen per 60 samples and 1 pathogen per 3.4 L water. John and Howard found the highest percentage of pathogens during spring and fall, whereas Kyle and Noblet (1987) found summer and fall to be the peak periods.

Acanthamoeba spp. have been occasionally detected in marine water and sediments. Most studies on *Acanthamoeba* spp. in marine sediments have been carried out in areas where sewage and other wastes have been dumped at sea (O'Malley et al. 1982; Sawyer et al. 1982). Sawyer et al. (1992) later recovered several species of *Acanthamoeba* from sewage-contaminated inshore New York and New Jersey shellfish beds that were periodically closed to shellfish harvesting. Munson (1993) recovered several species of *Acanthamoeba* from coastal waters of Bermuda, noting a high frequency of recovery of *Acanthamoeba* spp. near sewage outfalls.

B. Tapwater and Bottled Water

Acanthamoebae have been detected in tapwater, and several studies have documented their occurrence; however, all these studies have been done in countries other than the U.S. Rivera et al. (1979) collected 25 one-gallon (1-gal) water samples from faucets in private residences in Mexico: flagellates were found in 84% of the samples, amoebae in 13%, and ciliates in 1.9%. Although found infrequently, *Acanthamoeba astronyxis* and *A. castellanii* were recovered from the same samples. In another study, Hamadto et al. (1993) tested 50 tap water samples in Egypt and recovered unidentified species of *Acanthamoeba* from 2 samples. Michel et al. (1998) tested drinking water in a new hospital in Germany and found amoebae in 20 of 37 (54%) samples; 2 of 16 isolates of *Acanthamoeba* were pathogenic to mice. Rohr et al. (1998) collected water from 56 hot water taps in hospitals, also in Germany, and found amoebae in 29 (56%) of them. The authors recovered five genera of cyst-forming amoebae but none were species of *Acanthamoeba*. In England, Seal et al. (1992) isolated *Acanthamoeba* from five of six bathroom cold water taps supplied by storage tanks and one kitchen cold water tap supplied by the mains. When 41 strains of amoebae were recovered from 49 swab samples collected from moist areas in the hospital, such as walls, floor tiles, and sinks, 22% were species of *Acanthamoeba*. In Germany, Michel et al. (1998) recovered a species of *Acanthamoeba* from a hospital cold water tap. In Hong Kong, Houang et al. (2001) found that 8% of the homes were colonized with *Acanthamoeba*.

The common occurrence of *Acanthamoeba* in eyewash stations filled with tapwater containing free chlorine (concentration of chlorine was not reported) has been reported in the U.S. (Bowman et al. 1996). *Acanthamoeba* are able to grow in stagnant water in eyewash stations, and regular flushing is required to control their numbers. The presence of free chlorine or other disinfectants was not reported in any of the previous studies.

Rivera et al. (1981) tested three popular brands of bottled mineral waters available in local stores in Mexico and identified *Naegleria gruberi*, *Vahlkamp-*

fia vahlkampfi, and *Acanthamoeba astronyxis*. The author did not state how or if the water had received any processing before bottling.

C. Swimming Pools and Spas

Residential and public pools and spas have been documented as frequent sources of amphizoic amoebae, including *Acanthamoeba*. When amoebae were first identified as a cause of meningitis, Lyons and Kapur (1977) tested water from 30 public pools in New York disinfected with either chlorine or bromine and recovered amoebae from 27 pools. The species were not identified but were referred to as belonging to the "*Hartmannella-Acanthamoeba*" group, a term often used before the two genera were recognized as distinct taxonomic entities. *Acanthamoeba* has been isolated in swimming pools or other bodies of water around the world, including Germany (Janitschke et al. 1980), Mexico (Rivera et al. 1983), and frozen swimming areas in Norway (Brown and Cursons 1977).

Thermal bathing pools (spas) are also sources for potentially pathogenic amoebae (Martinez 1985). Brown et al. (1983) tested nine thermal pools in New Zealand and identified temperature-tolerant strains of *Acanthamoeba* from 20 % of them. They set up 88 subsamples from the pools and found *Acanthamoeba* in 5 (5.7%). Rivera et al. (1987) studied three resorts in Mexico that received water flowing from natural springs of thermal water. They recovered 12 strains of *Acanthamoeba* from cultures incubated at 42 °–45 °C; 2 strains were identified as *A. castellanii*, 1 as *A. lugdunensis*, and the others as *Acanthamoeba* spp. All were pathogenic to mice. In a second study (Rivera et al. 1991) they recovered *A. culbertsoni* and *A. polyphaga* from heated physiotherapy tubs. Penas-Ares et al. (1994) tested heated water used to fill 12 spas in Spain; the water was classified as sulfurous, and temperatures ranged from 34° to 64 °C. The authors recovered 13 strains of amoebae from 8 of the spas. Four of the 8 spas yielded *A. polyphaga* or *A. lenticulata*, but only *A. polyphaga* was found to be pathogenic to mice. The amoebae may survive pool and spa disinfection procedures because of their resistant cyst stages.

D. Sewage and Biosolids

Daggett (1982) published a description of potentially pathogenic *Acanthamoeba* and *Naegleria* in polluted waters with emphasis on health risks to divers. Singh and Das (1972) studied biosolid samples in Bombay, India, and recovered strains of *Acanthamoeba culbertsoni* and *A. rhysodes* that were pathogenic to mice. Bose et al. (1990) extended studies on sewage in India to include Calcutta, where they isolated a pathogenic strain of *A. castellanii* and a nonpathogenic strain of *A. astronyxis*.

E. Animal Wastes

Bovee et al. (1961) tested intestinal contents from reptiles in Florida using the agar plate method and recovered amoebae from 35 of 157 fecal samples. Wilson et al. (1967) conducted a follow-up study in Florida, identifying cyst-forming genera of amoebae representing *Acanthamoeba* from water and the intestinal contents of snakes and lizards. Jadin et al. (1973) carried out an extensive study on wildlife in France and recovered *Acanthamoeba* from the feces of snakes, toads, frogs, ducks, gulls, and muskrats. The study showed that animals largely aquatic in habitat could be sources of *Acanthamoeba* in natural bodies of water. Franke and Mackiewicz (1982) discovered animals that transport *Acanthamoeba* in their feces by culturing *A. polyphaga* from the common shiner *Notropis cornatus* and the white sucker *Catostomies commersari* from streams in New York. Simitzis-LeFlohic and Chastel (1982) reported finding species of *Acanthamoeba* in feces of small feral mammals in Brittany, Tunisia, and France.

F. Air, Dust, and Soil

Air is a carrier of dust, dirt, fungal spores, and other forms of particulate matter. During a dust storm in Zaire, Africa, Lawande et al. (1979) collected nasal swabs from 50 children ranging in age from 1 mon to 10 yr and recovered soil amoebae from 12 (24%). Two of the 12 children harbored *A. rhisodes*. Lawande et al. (1979) also exposed open culture plates to the atmosphere for periods of 30 min to 4 hr. Amoebae identified as *A. castellanii* and *A. culbertsoni* were recovered as early as 30 min after the plates were opened. The study throughout the 4-hr period yielded other species as well, including *A. astronyxis, A. palestinensis*, and *A. rhisodes*. Rivera et al. (1987) conducted similar studies during the rainy season in Mexico City, Mexico, recovering *A. astronyxis, A. castellanii, A. culbertsoni*, and *A. polyphaga* from air. In a second study of air in Mexico, Rivera et al. (1991) recovered nine species of *Acanthamoeba*. Air conditioners and cooling towers also contribute moisture and microbial pathogens, including *Acanthamoeba*, to the atmosphere (Walker et al. 1986; Ma et al. 1990; el Sibae 1993). In quantitative studies on the density of *Acanthamoeba* cysts in outdoor air, Kingston and Warhurst (1969) recorded values of one cyst/m^3 and one cyst of *A. castellanii*/18.3 m^3 of air.

In summary, *Acanthamoeba* can be isolated from most aquatic environments, air, and soil. Their concentration in water is related to the number of bacteria upon which they feed. Little quantitative information is available on their concentration in water, and their occurrence in distribution systems and tapwater has not been systematically studied in the U.S. Recreational exposure may occur because of their presence in swimming pools, hot tubs, and surface waters. They may occur seasonally in greater numbers in the early spring and early fall.

IV. Health Effects

Two types of illness are most commonly associated with *Acanthamoeba* spp., *Acanthamoeba* keratitis (an infection of the eye) and granulomatous amoebic encephalitis (GAE). GAE infection is usually considered opportunistic. Keratitis occurs primarily in healthy individuals who wear contact lenses and GAE occurs primarily in immunodeficient individuals. A comparison of the clinical and pathological features of the two diseases is listed in Table 6.

Table 6. Comparison of clinical and pathological features of granulomatous amoebic encephalitis (GAE) and *Acanthamoeba* keratitis (AK).

Features	GAE	AK
Predisposing factors	Immunodeficiency; AIDS; debilitating chronic disease	Good health, corneal trauma, wearing of contaminated contact lens
Epidemiology	Worldwide	Worldwide
Usual portals of entry	Lungs; skin; nose; neuro-epithelium	Corneal abrasion
Incubation period	Probably weeks to months	Probably days
Clinical course	Subacute or chronic (several weeks to months)	Subacute or chronic
Prognosis	Almost always fatal	Good if properly treated
Clinical symptoms and signs	Personality changes; confusion; seizures; nausea; headache; dizziness	Eye pain; typical corneal ring "infiltrate"; photophobia; blurred vision
Treatment	Itraconazole; Miconazole; Sulfametazine; Pentamidine IV (in vitro)	Polyhexamethylene biguamide; propamidine isethionate

Risk of acanthamoebic eye infection is associated with physical injury to the eye or wearing of contact lenses in conjunction with exposure to water containing *Acanthamoeba* such as tapwater, hot tubs, natural springs, bottled water, and nonsterile water used to store contact lenses. Reports indicate that 85% of cases are associated with individuals who wear contact lenses.

GAE is a chronic illness of the central nervous system that affects the brain and is associated with *Acanthamoeba* spp. It is an infection primarily of immunocompromised individuals that usually leads to death.

A. Eye Infections (Acanthamoebic Keratitis)

Acanthamoeba species cause acanthamoebic keratitis, a painful, vision-threatening disease of the cornea. The infection is associated with minor corneal trauma or the use of contact lenses in normal, healthy people. Males and females are equally affected. *Acanthamoeba* keratitis is characterized by severe ocular pain, a complete or partial paracentral stromal ring infiltrate, recurrent corneal breakdown of the epithelium, and a corneal lesion refractory to commonly used ophthalmic antibacterial medication. Clinical features of the disease are listed in Table 7.

Some species of *Acanthamoeba* were not found to be associated with eye disease until the early 1970s. Jones et al. (1973, 1975) and Visvesvara et al. (1975) described the case of a rancher who scraped his eye while bailing hay and rinsed it with tap water pumped into his house from a well that used unfiltered river water. The authors also described an infection in a young female

Table 7. Characteristics and symptoms of patients with *Acanthamoeba* keratitis.

Young, healthy individuals
Wearers of soft contact lens
Nonpreserved or nonsterile solution used for storage of contact lens
Eye trauma
Usually one eye affected
Extreme eye pain
Corneal breakdown of the epithelial
Late in the infection, a corneal ring infiltrate is seen

---

nurse who had no history of eye disease, and a fatal infection in a 7-yr-old boy who had played in drainage ditches near his home. Nagington et al. (1974) described an eye infection in a 32-yr-old schoolteacher who did not have a history of exposure to contaminated water, and a second fatal case in a 59-yr-old farmer who was hit in the eye by a tree branch. Jones et al. (1975) also described a case involving a 58-yr-old farmer who had been exposed to dust while bailing barley on his farm. The infection failed to respond to treatment, and the eye had to be surgically removed.

Other cases of physical damage include irritation by an insect (Hamburg and de Jonckheere 1980), contamination by barley dust (Jones et al. 1975), and wind-surfing (Volker-Dieben et al. 1980). The effects from eye trauma ranged from successful treatment to corneal replacement, loss of the affected eye, and, rarely, death of the patient. Jones et al. (1975) described a fatal case in a young boy who was suspected of playing in a watering trough for cattle.

Eye infections reported in the 1970s generally were unique case histories involving injury. All of this changed when some of the eye infections thought to be of viral origin were found to be caused by *Acanthamoeba* (MMWR 1987). Ormerod and Smith (1986) reviewed the histories of 42 cases of keratitis in California that occurred between 1977 and 1984 and suggested that it was likely that extended-wear lenses might increase the risk of microbial keratitis. Stehr-Greene et al. (1987) conducted a case-control study to obtain information on the role of contact lens sanitary practices on injury to the eye. They studied 27 patients with keratitis and 81 uninfected individuals (controls) to compare lens care practices. Patients with keratitis were found more likely to use homemade solutions than controls (78% vs. 17%) and were more likely to wear lenses while swimming (63% vs. 30%). The authors found that microbial contaminants other than *Acanthamoeba* were present in 1 of 59 commercial saline solutions, 11 of 11 homemade solutions, and 23 of 29 bottles of nonsterile distilled water. Thus, there is little doubt that microorganisms in nonsterile cleansing solutions may become established in contact lens cases, perhaps on the lenses themselves, and lead to serious eye disease. Badendoch (1991) and Martinez and Visvesvara (1997) have reviewed most of the literature on amoebic eye diseases beginning with some of the earliest recognized cases and noted that successful outcomes depended on early diagnosis and treatment. Martinez and Visvesvara (1997) estimated that, as of January 1996, more than 750 cases of amoebic keratitis had been reported worldwide.

There are several important risk factors associated with acanthamoebic keratitis. Most patients have at least one of these identifiable factors, which include corneal trauma, exposure to contaminated water, and contact lens use. Approximately 71%–85% of patients with acanthamoebic keratitis are contact lens wearers (Moore and McCulley 1989; Moore et al. 1985).

No single type of contact lens has been excluded from association with acanthamoebic keratitis. People with daily-wear soft contact lenses account for approximately 75% of the cases, those with extended-wear contact lenses about 14%, with hard contact lenses about 6%, and people with rigid gas-permeable lenses account for about 4% (Moore et al. 1985). In another study, Stehr-Green et al. (1987) reported that most patients (95%) had at least one risk factor for acanthamoebic keratitis; of the 85% who wore contact lenses, most wore daily-wear (56%) or extended-wear (19%) soft lenses. Some patients (including both types of contact lens wearers, 26%) had a history of corneal trauma before developing acanthamoebic keratitis, and 25% of patients had a history of exposure to contaminated water.

Two studies have identified tapwater washing of lens cases in cases of *Acanthamoeba* (Seal et al. 1996, Ledee et al. 1996). Using molecular fingerprinting techniques, Ledee et al. (1996) established domestic tapwater in the United Kingdom as the source of contamination in contact lens wearers. Similarly, contact lens wearers who have been exposed frequently to hot tubs or natural springs are at risk of developing acanthamoebic keratitis (Wilhelmus and Jones 1991).

Symptoms of Acanthamoeba *keratitis*   Clinical symptoms are usually a history of pain and the formation of a whitish halo or ring infiltrate around the periphery of the cornea (Fig. 3). Although most cases present a history of contact lens wear, the infections are also associated with a foreign object or physical trauma in the affected eye.

Diagnosis of Acanthamoeba *Keratitis*   Positive diagnosis of acanthamoebic keratitis can be made by *in vivo* confocal microscopy, but diagnostic tests usually rely on demonstrating amoebae on corneal scrapings or biopsy material (Seal et al. 1996). Samples of corneal epithelium and any infiltrated stroma are removed under local anesthetic, and contact lenses and storage cases may also be cultured. The most common method is to inoculate the sample into the center of a nonnutrient agar plate seeded with *E. coli* (Singh and Petri 2000). With incubation at 32 °C in air, migration tracks are usually visible within 48 hr. Positive identification requires some experience, and it is useful to incubate a control plate that is not inoculated with a clinical specimen.

Identification Procedures   Standard methods for morphological characterization, isoenzyme electrophoresis, immunological techniques, and temperature tol-

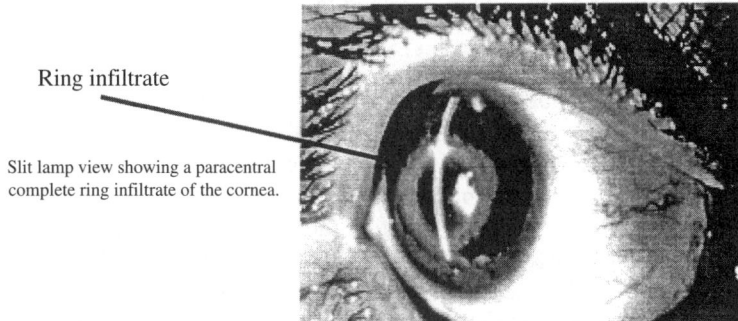

Ring infiltrate

Slit lamp view showing a paracentral complete ring infiltrate of the cornea.

Infected Eye showing the diagnostic concentric ring

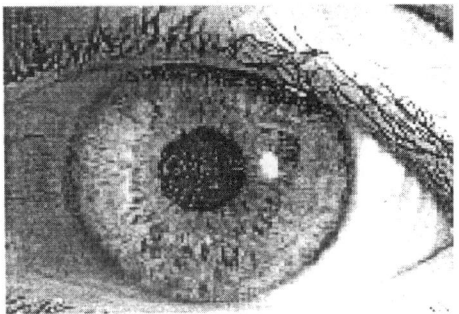

Normal Eye

Fig. 3. Appearance of cornea in *Acanthamoeba* keratitis. (Top figure), reprinted from Ophthalmology, 92:1471–1479, Theodore FH et al: "The diagnostic value of a ring infiltrate in *Acanthamoeba* keratitis." © 1985 American Academy of Ophthalmology.

erance tests have been published and widely used (Singh and Petri 2000). Results obtained by using one or more of these techniques, coupled with animal pathogenicity tests and the shape and size of cysts, are often adequate for identifying more commonly occurring species of *Acanthamoeba*.

Corneal biopsy of the infected eye is usually sufficient for confirming infection by amphizoic amoebae. However, it may be possible to make an identification of genus when distinctive double-walled wrinkled cysts suggest a Group III species of *Acanthamoeba*. When amoebae from corresponding pieces of tissue appear on culture plates, the cysts are often distinctive enough to place the organism in *Acanthamoeba*. Keys to soil amoebae (Page 1976, 1988) or photographs (Pussard and Pons 1977) often are sufficient for identifying some of the well-known species. Biochemical methods for obtaining isoenzyme profiles (de Jonckheere and Michel 1988) are extremely useful in combination with morphological features for identifying most amoebae (Sawyer 1992). Griffin

(1972) used thermotolerance as one method for screening amoebae for pathogenicity.

Treatment of Acanthamoebic Keratitis In the first 10 yr after the emergence of acanthamoebic keratitis as a clinical problem, treatment was usually unsatisfactory, employing a wide variety of topical agents in combination. In 1985, Wright et al. reported successful medical treatment using propamidine isethionate (Brolene) 0.1%, an aromatic diamidine, applied topically with dibromopropamidine ointment 0.15%, and followed by treatment with neomycin when signs of toxicity occurred. The success of the treatment was attributed to the amoebicidal activity of both propamidine and dibromopropamidine, although subsequently dibromopropamidine was generally omitted from the regimen. Further experience showed that a medical cure with propamidine therapy was most likely to be achieved if treatment began early in the course of the disease (Moore and McCulley 1989). Propamidine was generally combined with neomycin, initially instilled hourly and tapered slowly over several months after improvement was noted. However, in some patients results were still poor, and more effective compounds were sought (Ficker 1988). Successful treatment using propamidine with miconazole 1% (often with neomycin sulfate-polymixin B sulfate-gramicidin) has been reported (Berger et al. 1990), as has combination therapy with oral itraconazole, with topical miconazole 0.1% and debridement (Ishibashi et al. 1990). Another combination regimen is topical clotrimazole 1%–2% with propamidine and neomycin sulfate-polymixin B sulfate-gramicidin; in one series, medical cure was achieved in 11 of 14 patients with eye infections using this combination (D'Aversa et al. 1995).

In the early 1990s, *in vitro* sensitivity studies showed that the cationic disinfectant polyhexamethylene biguanide (PHMB) was highly effective in killing both cysts and trophozoites, and in 1992 Larkin et al. reported its successful clinical use at a concentration of 0.02%. The main theoretical advantage of PHMB over other compounds seems to be its consistently high cysticidal activity against a number of strains, compared with other compounds that may be active against some strains but relatively ineffective against others. Another factor is that, in contrast to propamidine, PHMB does not appear to be associated with toxicity problems (Johns et al. 1988). Clinical experience with PHMB, usually in combination with propamidine, has shown that if used early enough in the course of the disease the prognosis is very good, and penetrating keratoplasty is unlikely to be necessary (Illingworth et al. 1995). Use of the diamidine derivative hexamidine, which appears to have a greater cysticidal activity than propamidine, was reported by Brasseur et al. (1994). The use of chlorohexidine 0.02% as an alternative to PHMB has also been reported, resulting in a medical cure in 11 of 12 patients (Seal et al. 1996).

Incidence of Acanthamoeba *Keratitis* Acanthamoeba keratitis is not a reportable disease in the U.S., so the true incidence is not known. Published work suggests an incidence of 0.58–0.71 cases/1,000,000 in the general population, and 1.65–2.01/1,000,000 among contact lens wearers (Schaumberg et al. 1998). One study in the UK reported an incidence of 149/1,000,000 among contact lens wearers (Seal 2000). Studies reporting the incidence of *Acanthamoeba* keratitis are summarized in Table 8. The incidence of all causes of microbial keratitis (largely bacterial) is about 400/1,000,000 among contact lens wearers. Worldwide, the incidence of microbial keratitis has been reported to range from 1.1/1,000,000 to 2,000/1,000,000 among contact lens wearers (Cheng et al. 1999). Difficulties in the diagnosis of *Acanthamoeba* keratitis probably lead to an underestimation of the true number of cases. An estimate of *Acanthamoeba* keratitis known cases in the U.S. at present is 500, with more than 3,000 cases worldwide (Martinez and Visvesvara 2001).

Pathogenicity The pathogenesis of acanthamoebic keratitis has been suggested to follow two pathways (Alizadeh et al. 1995). The first pathway is restricted to the epithelium without involvement of the stoma and has a good prognosis. The second pathway culminates in the parasites entering the stoma, resulting in extensive necrosis and edema. The first step in the initiation of infection is the attachment to the epithelial surface. Amoebae bind to the corneal surface and produce epithelial thinning and necrosis.

The pathogenicity of *Acanthamoeba* spp. is related to its ability to attach to corneal epithelial cells. Khan (2001) found that *Acanthamoeba* exhibited higher number of acantodia, structures associated with the binding of amoeba to the target cells in the eye, as compared to nonpathogenic *Acanthamoeba*. Additional results indicated that phagocytosis occurs in the pathogenic amoeba by formation of amoebastone, characteristic of amoeba phagocytes, and that *Acanthamoeba* phageocytosis may be both an efficient means of obtaining nutrients and a significant factor in pathogenesis of *Acanthamoeba* infections.

Table 8. Worldwide incidence of *Acanthamoeba* keratitis.

Incidence per 1,000,000	Population	Country	Year(s)	Reference
1.65–2.01	CLW	USA	1985–1987	Schaumberg et al. (1998)
1.1	CLW	Netherlands	1996	Cheng et al. (1999)
149	CLW	UK	1996	Seal (2000)
0.58–0.71	GP	USA	1985–1987	Schaumberg et al. (1998)
1.40	GP	UK	1996	Radford et al. (1998)
1.30	GP (Iowa well water)	USA	1993–1994	Meier et al. (1998)
14.3	GP (during flooding of municipal systems)	USA	1993–1994	Meier et al. (1998)

CLW: contact lens wearer; GP: general population.

Khan et al. (2001) differentiated pathogenic *Acanthamoeba* by their ability to produce cytopathogenic effects (CPE) on corneal epithelial cells in culture. They also reported that pathogenic *Acanthamoeba* showed growth on higher osmolarity (1 M mannitol) whereas growth of nonpathogens was inhibited. The pathogenic potential of *A. castellani* isolates was correlated with the ability to bind to the corneal epithelium, respond chemotactically to corneal endothelial extracts, elaborate plasminogen activators, and produce cytopathogenic extracts (van Klink et al. 1992).

The 18S rRNA gene (Rns) phylogeny of *Acanthamoeba* has been investigated as a basis for improvements in the nomenclature and taxonomy of the genus (Stothard et al. 1998). Twelve linages referred to as T1–T12 have been identified, with most of the keratitis-causing strains belonging to group T4 (Stothard et al. 1998; Walochnik et al. 2000). More recently, type T6 has also been reported to be associated with keratitis (Walochnik et al. 2000).

Another factor in the pathogenicity of *Acanthamoeba* may be an individual's ability to produce antibodies in tears (Alizadeh et al. 2001). The presence of serum antibody in 50%–100% of the population suggests that exposure to *Acanthamoeba* species is ubiquitous (Cursons et al. 1980b; Cerva 1989). However, patients with *Acanthamoeba* keratitis have significantly higher anti-*Acanthamoeba* IgG antibody titers than heathy subjects (Alizadeh et al. 2001). In contrast, anti-*Acanthamoeba* tear IgA was significantly lower in patients with *Acanthamoeba* keratitis in comparison with healthy subjects, suggesting that a low level of anti-*Acanthamoeba* IgA antibody in the tears may be associated with *Acanthamoeba* keratitis.

In summary, the pathogenic potential of *Acanthamoeba* appears to be related to certain strains and the ability of the host to produce IgA antibodies in the tears.

Immunity The presence of serum antibody in 50%–100% of the population suggests that exposure to *Acanthamoeba* species is common (Cursons et al. 1980a; Cerva 1989). These antibodies were shown to be capable of neutralizing cytopathogenic effects of *Acanthamoeba* (Ferrante 1991). Patients with *Acanthamoeba* keratitis have a significantly higher anti-*Acanthamoeba* IgG antibody titer than healthy subjects (Alizadeh et al. 2001). In contrast, anti-*Acanthamoeba* tear IgA is significantly lower in patients with *Acanthamoeba* keratitis in comparison with healthy subjects, suggesting a low level of anti-*Acanthamoeba* IgA antibody in the tears may be associated with *Acanthamoeba* keratitis. Persistent corneal and scleral inflammation observed following cases of *Acanthamoeba* keratitis is not always caused by active amoebic infection but can be due to persisting acanthamoebic antigens. Yang et al. (2001) found that *Acanthamoeba* cysts can persist for up to 31 mon in the eye after treatment although trophozoites are no longer present. They hypothesized that *Acanthamoeba* cysts can remain in corneal tissue for extended periods of time and may cause persistent inflammation in the absence of active amoebic infection.

The feasibility of inducing protective immunity to *Acanthamoeba* keratitis has been tested in a pig model (Alizadeh et al. 1995). It was possible to induce immunity in 50% of the animals by subconjunctival injection of the parasites, and in 100% by a combination of intramuscular and subconjunctival injection, whereas corneal infection alone did not confer immunity to subsequent infection.

B. Granulomatous Amoebic Encephalitis

Granulomatous amoebic encephalitis (GAE) caused by *Acanthamoeba* spp. is the second major infection associated with *Acanthamoeba*. GAE is a chronic, progressive disease of the central nervous system occurring most often in persons with poor immune systems or other debilitating health problems. Predisposing factors include chemotherapy, dialysis, diabetes mellitus, treatment with steroids, chronic alcoholism, smoking, bone marrow or renal transplantation, or acquired immunodeficiency syndrome (AIDS) (Marciano-Cabral et al. 2000). Chronic skin infections have been reported from patients with GAE. However, it is not known whether skin lesions provide the primary site of infection or represent terminal dissemination of *Acanthamoeba* from the lungs to other sites (Marciano-Cabral et al. 2000). In most AIDS patients, skin lesions and sinusitis, which may be caused by *A. astronyxis*, *A. palestinensis*, A. culbertsoni, or *A. castellanii*, are common features. The infection spreads from lung or skin lesions to the central nervous system, resulting in neurologica deficits that progress over days or weeks to meningoencephalitis and death.

Another free-living amoeba, *Naegleria fowleri*, was later discovered to cause an aseptic meningitis that was usually fatal (Ma et al. 1990). The term primary amoebic meningoencephalitis, or PAM, was proposed for infection by *Naegleria* (Butt 1966), and the term granulomatous amoebic encephalitis, or GAE, was proposed for infections by *Acanthamoeba* (Martinez 1980). The two disease entities differ because PAM occurs most often in young people, is associated with swimming, and has a rapid onset of symptoms. In contrast, GAE occurs most often in patients with poor immune systems or patients suffering from long-standing health problems regardless of age. GAE caused by *Acanthamoeba* or *Balamuthia* is now recognized as a disease occurring most often in persons with poor immune systems or suffering from some other debilitating health problem (e.g., alcoholism, diabetes, smoking, or AIDS). The amoebae are believed to enter the bloodstream, probably via the nose, lungs, or breaks in the skin following injury or trauma: they then affect various organs by hematogenous spread. *Balamuthia* has been identified in approximately 40 patients in the U.S., including more than 10 with AIDS infection (Martinez et al. 1997; Visvaresvara 2001). In contrast, *Acanthamoeba* has accounted for approximately 84 (~50 with AIDS) cases in the U.S. and 120 worldwide (Martinez et al. 1997; Visvaresvara 2001). The disease may be the end result of long-term injury. Fatal infections probably occur in individuals with extensive damage to the central nervous system and internal organs before the manifestation of overt clinical symptoms.

The exact pathway of amoebae entering the brain is difficult to determine, because in most cases with a fatal outcome there has been a history of predisposing factors. It is believed that the amoebae are spread throughout the body via blood vessels (hematogenous spread), after entry through the nasal passages, lower respiratory system, or breaks in the skin caused by injury (Ma et al. 1990). Patients who have been treated for GAE range from children to elderly adults with a clinical history of illness ranging from about 1 wk to 6 mon (Martinez et al. 1977). Symptoms of neurological disease on admission to a hospital are varied, including headache, drowsiness, low-grade fever, and stiffness of the neck. Other symptoms that may appear early in the disease are personality changes, seizures, nausea, vomiting, or lethargy (Martinez and Visvesvara 1991). Thorough diagnostic procedures are necessary to recognize amoebic meningoencephalitis because on initial examination the disease is not always easy to distinguish from bacterial meningitis, tuberculous meningitis, brain tumors, or viral meningitis (Martinez and Visvesvara 1997). Martinez and Janitschke (1985) reviewed 33 cases of GAE and listed several illnesses associated with the patients who had the disease that included skin ulcers, cirrhosis of the liver, hepatitis, pneumonitis, renal failure, collagen-connective tissue disease, and pharyngitis. Predisposing factors mentioned by the authors included chemotherapy, radiation treatment, steroids, broad-spectrum antibiotics, alcoholism, splenectomy, and peritoneal dialysis.

Diagnosis and Treatment of GAE Patients with confirmed GAE usually are chronically ill, immunosuppressed, or debilitated by other causes. By the time a diagnosis has been made, the central nervous system may have been invaded, probably via the nasal passages, respiratory tract, or skin (Martinez 1993). The diagnosis may be questionable at first because of the possibility of brain tumor, abscess, or intracerebral hematoma (Visvesvara et al. 1997). Successful treatment is rare, and infection usually results in the death of the patient. *In vitro* studies have shown that diamidine derivatives such as pentamidine, propamidine, miconazole, ketoconazole, and 5-fluorocytosine may be effective in treating GAE (Martinez et al. 1997). Sometimes skin nodules harboring *Acanthamoeba* are detected before spreading to internal organs and the central nervous system. Visvesvara et al. (1997) suggested that when skin nodules or ulcers are present, treatment may be tried using topical chlorhexidine gluconate and intravenous pentamidine.

In spite of the poor prognosis for most patients with GAE, efforts to find at least a partially successful treatment are in progress. A new class of peptide compounds called magainins may have amoebostatic and amoebicidal properties when used with other amoebicidal agents (Martinez et al. 1997; Schuster and Jacob 1992). Schuster and Visvesvara (1998) tested antimicrobials and phenothiazine compounds against amphizoic amoebae but found the levels affecting them probably were too high for clinical use. In other efforts, Chu et al. (1998) have studied the effects of plant extracts that were amoebicidal or induced encystment.

Incidence of GAE The global incidence as of 2000 stood at 120 recorded GAE cases; 84 of those occurred in the U.S., and more than 50 of the GAE cases were found in AIDS patients (Martinez and Visvesvara 2001). There is general agreement that both GAE and keratitis have increased in the past 10 years in the U.S. because of the increase in contact lens wearers of all ages for various reasons, including athletic and cosmetic, and the increase in the number of immunosuppressed individuals (Marciano-Cabral et al. 2000; EPA 1998).

Pathogenesis and Immunity The pathogenesis of GAE is complex and poorly understood (Martinez and Visvesvara 1997). In GAE, the immunity is predominantly T-cell mediated; therefore, the diminution of CD+ and T helper lymphocytes, as occurs in AIDS patients, enables the proliferation of free-living amebae. Ulceration of the skin containing both amebic trophozoites and cysts suggests also the portal of entry into the bloodstream. In experimental animals, the olfactory neuroepithelium has also been found to be a possible portal of entry (Janitschke et al. 1996). The incubation period of GAE is unknown but is probably longer than 10 d. The ability of the *Acanthamoeba* to produce necrosis of the brain tissue is probably due to an enzymatic action induced by lysosomal hydrolases and phospholipase that can degrade the phospholipids of the myelin sheaths (Martinez and Visvesvara 1997).

Studies in mice have demonstrated that it is possible to immunize animals against *Acanthamoeba* meningoencephalitis (Culbertson 1971; Rowan-Kelly and Ferrante 1984). Animals immunized intraperitoneally with sonicated trophozoites of *A. culbertsoni* were highly resistant to intranasal infection with the organism; however, those immunized with a nonpathogenic *A. culbertsoni* or *A. polyphaga* were not protected against infection with *A. culbertsoni*.

GAE in Domestic Animals and Wildlife Several reports of amphizoic amoebae in animals appeared in the literature at about the same time as they were found in fatal infections in humans. The principal difference between human and animal infection is that infection in humans occurs primarily in persons with deficient immune systems or those taking immunosuppressive drugs, which is not found in cases involving animals. Kadlec (1978) carried out one of the most extensive surveys of infection in domestic animals by amphizoic amoebae that identified *Acanthamoeba* spp. from bulls, cows, a rabbit, pigeons, and turkeys. Infection in animals probably occurs by the same routes as reported for humans. It has also been described in dogs by several investigators (Ayers et al. 1972; Bauer et al. 1993). Infections in the lung of water buffalo and bulls could have had a nasopharyngeal origin from drinking unclean water (Dwivedi and Singh 1965; McConnell et al. 1968). Evidence for water as a source of infection in animals by *Acanthamoeba* is found in reports of the amoebae in the gills, spleen, urinary bladder, or blood of wild-caught and ornamental fish (Taylor 1977; Dykova et al. 1996; Booton et al. 1999).

C. Other Infections Caused by Acanthamoeba

Occasional infections by *Acanthamoeba* spp. have included a purulent discharge from an ear (Lengy et al. 1971), a granulomatous skin lesion (Gullet et al. 1979), rhinosinusitis in an AIDS patient (Teknos et al. 2000) and possible association with intestinal disorders (Hoffler and Rubel 1974; Mehta and Guirges 1979; Thamprasert et al. 1993).

D. Immunocompromised Individuals

Several reports of *Acanthamoeba* infection in AIDS patients have involved the skin, as well as other tissues, and in most cases there was a fatal outcome in spite of treatment. In AIDS patients it is not always absolutely clear whether the AIDS virus or the amoeba was the primary cause of death. The infection with free-living amoebae is a terminal event. Individuals with deficient immune systems, whether natural or acquired, represent a segment of the population that is most likely to succumb to infections with microbial pathogens, including amphizoic amoebae. Gonzalez et al. (1986) reported a case resulting in death in a 29-yr-old patient with AIDS. At autopsy, amoebae were found in the paranasal sinuses, a calf nodule, and an abscess of the left leg, but not in the brain. The following year, Wiley et al. (1987) examined a 34-yr-old patient with a history of nasopharyngeal allergies and infections with *Giardia lamblia* and *Cryptosporidium* spp. The patient underwent an appendectomy and developed a hard skin nodule above the surgical scar. The patient stated that he had noticed painful skin lesions before surgery. At autopsy, amoebae were found in the brain and the skin. Tissue fragments placed in kidney cell tissue cultures yielded amoebae identified as *Acanthamoeba culbertsoni*. Another case involving skin infection was reported by Friedland et al. (1992), who treated an AIDS-infected 8-yr-old Hispanic boy who died of the infection. The patient had a persistent nasal discharge and skin nodules that eventually became ulcerated and 2–4 mm deep before death. Gordon et al. (1992) described a fatal case in an AIDS patient caused by *A. polyphaga*, and Gardner et al. (1991) described a case probably caused by *A. rhisodes*. Other fatal cases in AIDS patients were reported in 1994 (Park et al. 1994) and 1996 (Telang et al. 1996).

Visvesvara et al. (1983) described a fatal case of GAE that involved a patient with a liver transplant. Twenty-six days after the transplant, the patient was readmitted to the hospital with pneumonia and cytomegalovirus infection. At autopsy, amoebae were found in the brain, lungs, blood vessel walls, adrenal and thyroid glands, lymph nodes, skin, and breast tissue. Borochovitz et al. (1981) identified *A. castellanii* from a bone graft in a diseased mandible. Anderlini et al. (1994) described two cases of fatal amoebic encephalitis in patients with leukemia who had received bone marrow transplants.

E. Effect of Endosymbiosis on Virulence

Acanthamoeba spp. have been demonstrated to develop endosymbiotic relationships with a number of waterborne bacteria, including *Legionella pneumophila* and *Mycobacterium avium* (see Table 4). This relationship may be important both in the growth and survival of these opportunistic pathogens in drinking water systems and in their ability to cause disease in humans.

Cirillo et al. (1997) found that *Mycobacterium avium* replicates within *Acanthamoeba castellanii* and that this association enhanced both entry and intracellular replication compared to the growth of the bacteria in broth culture. Furthermore, amoeba-grown *M. avium* was also more virulent in a mouse model. They also found that the highest growth rate of the *M. avium* in the amoebae was near 37 °C. From this observation, they suggested that if growth of *M. avium* in water environments occurs primarily within protozoa, the fact that *M. avium* has temperature-dependent growth in amoebae may explain why *M. avium* infections are more frequently associated with warm water supplies. It was also found that nonpathogenic strains of *Mycobacterium* were readily killed within the amoeba.

Cirillo et al. (1999) found *Legionella pneumophila* grown in *A. castellanii* to be at least 100 fold more invasive for macrophages than when grown on agar. They also provided evidence that amoeba-grown *L. pneumophila* expressed different proteins that may have been related to its enhanced invasiveness. The authors also suggested the replication of *L. pneumophila* in protozoans present in domestic water supplies may be necessary to produce bacteria that are competent to enter mammalian cells and produce human disease. A recent study has suggested that endosymbiosis enhances the virulence of the *Acanthamoeba*. Fritsche et al. (1998) reported that endosymbiont-infected amoebae produced a statistically significant enhancement in cellular destruction of human embryonic tonsilar (HET) cell monolayers in comparison to uninfected amoeba. Neither the bacteria or *Acanthamoeba* alone was capable of producing cellular destruction (i.e., cytopathic effects). Whether such enhanced pathogenic effects occurs in clinical *Acanthamoeba* infections is unknown.

V. Risk Assessment

Certain species of the genus *Acanthamoeba* have been associated with eye disease in humans. Five such species are listed in Table 9. Most of the infections (85%) in the U.S. are associated with the use of contact lenses and the remainder with some trauma to the eye (Stehr-Greene et al. 1987). Infection results from exposure to *Acanthamoeba* through improper storage of lenses, wetting of the lenses with unsterile solutions, improper disinfection of lenses, or swimming while wearing contact lenses. One epidemiological study suggests that increased risk may exist from municipal supplies that have been subjected to flooding (Meier et al. 1998). The concentration of free-living amoebae in surface waters may vary seasonally, creating a greater exposure at certain times of the year. *Acanthamoeba* is common in the aquatic environment (see Section III), and its cyst form is resistant to inactivation by chlorine (Radford et al. 1998). Wetting or storage of lenses in tapwater appears to be the most significant route of exposure for contact lens wearers. Molecular-based investigations have estab-

Table 9. Human infection caused by species of *Acanthamoeba*.

Species of *Acanthamoeba*	CNS infection	Eye infection	Other tissues	Reference
A. astronyxis	X		Adrenal, lymph node, sinus, skin, thyroid	Gullett et al. (1979)
A. castellanii	X	X	Lung, prostate, bone, muscle, sinus, skin	Martinez (1982) Martinez et al. (1977) Moore et al. (1985) Borochovitz et al. (1981) Gonzalez et al. (1986)
A. culbersoni	X	X	Liver, spleen, uterus, skin	Martinez et al. (1977) Wiley et al. (1987) Mannis et al. (1986) May et al. (1992)
A. divionensis	X			Di Gregorio et al. (1992)
A. griffini		X		Ledee et al. (1996)
A. hatchetti		X		Cohen et al. (1985)
A. healyi	X			Kim et al. (2000)
A. palestinensis	X			Ofori-Kwakye et al. (1986)
A. polyphaga		X		Singh and Petri (2000)
A. rhysodes	X	X		Singh and Petri (2000)

CNS: central nervous system.

lished domestic tapwater in the UK as a proven source of *Acanthamoeba* infection in lens wearers (Ledee et al. 1996). The organisms have been isolated from household taps and probably feed on the microbial biofilm within the distribution system.

An epidemiological study in the midwest U.S. suggested that an epidemic of presumed *Acanthamoeba* infections was associated with municipal water supplies subjected to flooding during 1993–1994 (Mathers et al. 1996; Meier et al. 1998). The incidence of presumed *Acanthamoeba* was 10 times greater (1.30 vs. 14.3 cases/10^6) in areas affected by flooding. The incidence was also significantly lower if the home was supplied with tapwater from a private well. In both these studies the authors used tandem scanning confocal microscopy and confirmatory cytopathological findings to diagnose the cases. However, the authors were unable to culture *Acanthamoeba* from individuals with keratitis. The authors suggested several reasons for their failure to culture the organism: (1) the infections were caused by a new species with different growth requirements; (2) the inoculum was insufficient; (3) an inhibitor was present; (4) the organisms were present but nonviable; or (5) the infections were caused by another organism.

A. Resistance to Drinking Water Treatment and Disinfection

No studies could be found on the effectiveness of drinking water treatment on the removal of *Acanthamoeba* cysts or trophozoites. Given the large size of the trophozoites (15–45 μm) and cysts (15–28 μm), they would be easily removed by filtration in a conventional water treatment plant. Their isolation from tapwater suggests that they can certainly colonize taps and feed on bacteria in the biofilm in distribution systems. de Jonckheere and Van de Voorde (1976) reported *Acanthamoeba* cysts to be very resistant to inactivation by chlorine, bromine, and iodine. The chlorine resistance of two different strains varied considerably. A 99.99% (4 $\log_{10}$) inactivation of a more sensitive strain was achieved with 16 mg/L within 1 hr. A 4 $\log_{10}$ decrease was not achieved after 24 hr with 6 mg/L.

The cysts have also been found to be very resistant to ultraviolet light. Chang et al. (1985) found the cysts of *A. castellanii* to be more resistant than *Bacillus subtilis* spores. A dose of approximately 70 mW-sec/cm^2 was required for a 99% (2 $\log_{10}$) inactivation of the cysts. The viability of the cysts was detected with a plaque assay on a lawn of *Escherichia coli* bacteria, requiring both excystation and growth of the organism.

In contrast, the trophozoites are much more sensitive to inactivation by chlorine and other disinfectants used to treat drinking water. A chlorine dose of 1.0 mg/L with a free chlorine residual of 0.25 mg/L after 30 min resulted in 99.99% reduction of trophozoites (Cursons et al. 1980a) of *A. castellanii* at pH 7.0 and 25 °C. A similar reduction with a chlorine dioxide dose of 2.9 mg/L (0.65 mg/L after 30 min) was achieved with chlorine dioxide and an ozone dose of 6.75 mg/L (residual, 0.078 mg/L after 30 min). The experiments were conducted in distilled water. Thus, although the trophozoites are inactivated by these disinfectants, they are significantly more resistant than bacteria. The resistance of *A. castellanii* to chlorine has been shown to add to the resistance of *Legionella pneumophila* growing within the *Acanthamoeba* and may play a significant role in the survival of opportunistic bacteria and their ecology and persistence in distribution systems, cooling towers, hot tubs, and other environments. Kilvington and Price (1990) reported that *A. polyphaga* were found to protect the legionellas from at least 50 mg/L free chlorine. Control of *Acanthamoeba* in distribution systems may be necessary for control of *Legionella pneumophila* and *Mycobacterium avium*.

B. Dose Response

Badenoch et al. (1990) demonstrated *Acanthamoeba* infections could be induced in the rat cornea by coinoculation with the bacterium *Corynebacterium xerosis*. Coinoculation with *C. xerosis* was necessary to induce the *Acanthamoeba* infection. Infection resulted in 7 of 24 rats that were exposed to 10^3 trophozoites and in 1 in 10 animals when exposed to 10^4 trophozoites. At least 10^4 *C. xerosis* had to be coinoculated to achieve these infection rates. The results suggest that at least 10^3 trophozoites are necessary to cause *Acanthamoeba* eye infection.

C. Risk Characterization

Acanthamoeba eye infections result from a combination of some eye trauma or contact lens use and other potential factors listed in Table 10. The concentration of *Acanthamoeba* in tapwater or aquatic environments may enhance the risk of infection. *Acanthamoeba* infections in contact lens wearers can be eliminated by proper care of the lens to avoid exposure to the organism. Exposure to contaminated water is the significant risk factor for contact lens wearers. Because *Acanthamoeba* cysts are resistant to inactivation by chlorine, a common disinfectant used for tapwater, exposure of the contact lenses to tapwater should be avoided. Proper disinfection of contact lenses and the solutions which they contact is essential to prevent infection.

Acanthamoeba may also play a significant role in the potential for transmission of *Legionella pneumophila* and *Mycobacterium avium* via drinking water. The growth of these organisms within *Acanthamoeba* may provide protection from disinfectants and enhance their ability to cause disease in humans. Providing an unsuitable habitat for *Acanthamoeba* could potentially reduce these risks. Low organic matter and disinfectant residuals would be expected to minimize the number of bacteria upon which the amoeba feeds. The amoeba population may also be limited in size, but not necessarily eliminated, by adequate disinfectant residuals.

Although it is clear that a relationship exists between *Acanthamoeba* in water and keratitis, the role of tapwater is not clearly understood. Data on the occurrence and concentration of *Acanthamoeba* in the United States are lacking. One study suggests that municipal supplies that may have become contaminated enhanced the risk of presumed *Acanthamoeba* keratitis (Meier et al. 1998). Seasonal distribution of keratitis and abundance of *Acanthamoeba* in surface waters also suggests a relationship. Additional information on the dose required for infection and quantitative data on occurrence in drinking water supplies would help to better understand the potential risks to contact lens wearers and the general public. The incidence of recognized *Acanthamoeba* keratitis is about 1–2/1,000,000 (see Table 8). The highest incidence in the U.S., which may have been connected to flooding and the use of municipal water supplies, was 14/1,000,000 (Meier et al. 1998). Even if all the cases of *Acanthamoeba* were associated with tapwater, this would be less than the 1:10,000 risk of infection

Table 10. Mechanisms involved in *Acanthamoeba* keratitis.

Previous epithelial trauma
Virulence of the organism
Number of organisms (on contact lens, in disinfection fluid, in contaminated water)
Capability of the ameba to adhere to the cornea
Duration of exposure
Immune response (presence of antibodies in tears)

per year that EPA has set as the goal for surface water supplies (EPA 1994; Regli et al. 1991).

VI. Association of Contact Lenses with Acanthamoebic Keratitis
A. Types of Contact Lenses

Contact lenses are worn on the surface of the eye to correct defects in an individual's vision. The first contact lens, made of glass, was developed in 1887 by Adolf Fick. The modern contact lens, which was developed in 1948, is made of plastic and rests on a cushion of tears (Table 11). It covers the cornea approximately over the iris and pupil. The hard plastic contact lenses had a limited wearing time because of potential irritation of the cornea. In the 1970s, soft lenses made from water-absorbing plastic gel for greater flexibility were introduced. In the 1980s extended-wear soft lenses, which can be worn without removal for several weeks at a time, were introduced. Soft contact lenses are usually more comfortable because they allow oxygen to penetrate to the surface of the eye. In the late 1970s, gas-permeable hard lenses, which allow more oxygen to reach the eye, were developed. The Food and Drug Administration must approve all contact lenses before they are available to the public. The types of contact lenses currently in use are listed in Table 12.

B. Demographics of Contact Lens Use

Currently it is estimated that 34 million Americans wear contact lenses (Contact Lens Council 2000). Approximately 85% of the wearers use soft contact lenses and 15% use the rigid gas-permeable type. Most wearers use daily-wear lenses that are removed at bedtime; 25% use extended-wear lenses (Table 13). Extended-wear lenses may be worn overnight and, in some cases, up to a week before removal. Only 13% of contact lens wearers are 17 years of age or younger (Table 14). Most soft contact lenses (45%) are worn by persons 26–39

Table 11. History of contact lens development.

Year	Event
1887	First contact lens made from glass; covers the entire eye
1939	Contact lenses first made from plastic
1948	Plastic contact lenses designed to cover the cornea only
1971	Introduction of soft contact lenses
1978	Introduction of oxygen-permeable lenses
1981	Food and Drug Administration approves soft contact lenses for extended (overnight) wear
1986	Overnight wear oxygen-permeable lenses become available
1987	Introduction of disposable soft contact lenses

Source: Contact Lens Council (2000) (*www.contactlenses.org*).

Table 12. Types of contact lenses.

Type	Comments
Daily-wear soft lenses	Made of soft, flexible plastics that allow oxygen to pass through to the eye. Cleaning is required.
Daily-wear disposable soft lenses	Typically no lens care is required
Extended-wear soft lenses	Available for overnight wear. Can usually be prescribed for up to 7 d of wear without removal.
Extended-wear disposable soft lenses	Worn 1–6 nights and then discarded. Require little or no cleaning.
Rigid gas-permeable lenses	Made of slightly flexible plastics that allow oxygen to pass through to the eye. Vision may be better than with soft lenses. Long life (1–2 yr). Daily and extended wear available.

Table 13. Wearers and types of contact lenses.

Type of lens	Percent of wearers
Soft lenses	85
Rigid gas permeable	15
Daily wear	75
Extended wear	25

Source: Contact Lens Council (2000).

Table 14. Age distribution of contact lens wearers in the United States.

Age (yr)	Percent of soft contact lens wearers	Percent of rigid gas-permeable contact lens wearers
<17	10	3
18–25	23	10
26–39	45	26
≥40	22	61

Source: Contact Lens Council (2000).

Table 15. Risk factors associated with acanthamoebic keratitis.

Risk Factor	Percent of Acanthamoebic keratitis cases
Wore contact lenses	85
Wore daily-wear lenses	56
Wore extended-wear lenses	19
History of corneal trauma	26
History of exposure to contaminated tapwater	25

years of age, In contrast, most rigid gas-permeable lenses are worn by persons 40 years and older.

C. Risk Factors

The use of contact lenses is the risk factor most commonly associated with acanthamoebic keratitis (Table 15). Stehr-Green et al. (1987) reported that 85% of the cases were associated with persons who wore contact lenses. All types of contact lenses have been associated with acanthamoebic keratitis (Table 16). Infection results from exposure to contaminated fluids used to wet the contact lens before placement on the eye or the use of contaminated fluids in storage cases. Any contact lens is a potential carrier of *Acanthamoeba* to the eye surface after being exposed to a contaminated fluid. The use of nonsterile solutions such as tapwater, bottled water and nonsterile distilled water has been associated with *Acanthamoeba* infections among contact lens wearers (Table 17) (Moore et al. 1985; Stehr-Green et al. 1987). Infection is also associated with wearing contact lenses during swimming (Stehr-Green et al. 1987), use of hot tubs, or exposure to natural springs (Wilhemus and Jones 1991). In a case-control study (MMWR 1987) it was found that of individuals who developed keratitis, 17 of 27 (63%) wore lenses while swimming whereas 24 of 81 (30%) did not. Also, patients with keratitis were more likely to wet lenses with saliva or wear lenses in a hot

Table 16. Types of contact lenses associated with acanthamoebic keratitis.

	Percentages of cases		
Type of contact lens	Illingworth et al. (1995)	Stehr-Green et al. (1987)	Moore et al. (1985)
Daily wear soft	21	56	75
Daily wear disposable soft	67	—	—
Extended wear	—	19	14
Hard	8	2	6
Rigid gas permeable	4	7	4

Table 17. Risk factors for acanthamoebic keratitis in contact lens wearers.

Use of tapwater to wet or store lenses
Use of bottled water to wet or store lenses
Use of distilled water to wet or store lenses
Use of nonsterile solutions to wet or store lenses
Wearing lenses during swimming
Wearing lenses in hot tubs
Wearing lenses in natural springs
Use of chlorine to disinfect lenses between uses
Wetting lenses with saliva

tub. The type of disinfectant used to treat the lenses during storage may also affect the risk of keratitis. Chlorine is not an effective means of disinfection and results in a greater risk of keratitis because of *Acanthamoeba* resistance to this disinfectant (Illingworth et al. 1995).

VII. Summary

Risk from *Acanthamoeba* keratitis is complex, depending upon the virulence of the particular strain, exposure, trauma, or other stress to the eye, and host immune response. Bacterial endosymbionts may also play a factor in the pathogenicity of *Acanthamoeba*. Which factor(s) may be the most important is not clear. The ability of the host to produce IgA antibodies in tears may be a significant factor. The immune response of the host is a significant risk factor for GAE infection. If so, then a certain subpopulation with an inability to produce IgA in the tears may be at greatest risk. There was no sufficient data on the occurrence or types of *Acanthamoeba* in tapwater in the U.S. Published work on amoebal presence in tapwater does not provide information on the type of treatment the water received or the level of residual chlorine. Assessment of the pathogenicity by cell culture and molecular methods of *Acanthamoeba* in tapwater would also be useful in the risk assessment process for drinking water. The possibility that *Acanthamoeba* spp. might serve as vectors for bacterial infections from water sources also should be explored. The bacterial endosymbionts include an interesting array of pathogens such as *Vibrio cholerae* and *Legionella pneumophila*, both of which are well recognized waterborne/water-based pathogens. Work is needed to determine if control of *Acanthamoeba* spp. is needed to control water-based pathogens in water supplies.

Acknowledgments

The authors thank the following individuals for the review of the Acanthamoeba health effects report and their helpful suggestions: Govinda Visvesvara, A. Julio Martinez, Walter Jakubowski, Jerry Niederkorn, Hassan Alizadeh, Hercules

Moura, Rita Schoeny, Paul S. Berger, Guy Carruthers, David Soderberg, Alfred DuFour, and James Sinclair.

References

Alizadeh H, He Y, McCulley JP, Ma D, Stewart GL, Via M, Haehling E, Niederkorn JY (1995) Successful immunization against *Acanthamoeba* keratitis in a pig model. Cornea 14:180–186.

Alizadeh H, Apte S, El-Agha MS, Li L, Hurt M, Howard K, Cavanagh HD, McCulley JP, Niederkorn JY (2001) Tear IgA and serum IgG antibodies against *Acanthamoeba* in patients with *Acanthamoeba* keratitis. Cornea 20:622–627.

Anderlini P, Przepiorka D, Luna M, Langford L, Andreeff M, Claxton D, Deisseroth, AB (1994) *Acanthamoeba* meningoencephalitis after bone marrow transplantation. Bone Marrow Transplant 14:459–461.

Asiri SMBA, Chinnis RJ, Banta WC (1990) Potentially pathogenic species of *Acanthamoeba* and *Hartmannella* (Protozoa: Amoebida) in sediment of the Potomac River near Washington, D.C. J Helminthol Soc Wash 57:88–90.

Ayers KM, Billups AH, Garner FM (1972) Acanthamoebiasis in a dog. Vet Pathol 9: 221–226.

Badenoch PR (1991) The pathogenesis of *Acanthamoeba* keratitis Australian and New Zealand J Ophthalmol 19:9–20.

Badenoch PR, Johnson AM, Christy PE, Coster DJ (1990) Pathogenicity of *Acanthamoeba* and a *Corynebacterium* in the. Arch Ophthalmol 108:107–112.

Bauer RW, Harrison LR, Watson CW, Styer EL, Chapman WL Jr (1993) Isolation of *Acanthamoeba* sp from a greyhound with pneumonia and granulomatous amebic encephalitis. J Vet Diagn Invest 5:386–391.

Berger ST, Mondino BJ, Hoft RH, Donzis PB, Holland GN, Farley MK, Levenson JE (1990) Successful medical management of *Acanthamoeba* keratitis. Am J Ophthalmol 110:395–403.

Booton GC, Dykova I, Lom J, Schroeder-Diedrich JM, Byers TJ (1999) Morphological and rDNA similarities of *Acanthamoeba* strains parasitic in fish and those causing human disease. J Eukaryo Microbiol 46:6A.

Borochovitz D, Martinez AJ, Patterson GT (1981) Osteomyelitis of a bone graft of the mandible with *Acanthamoeba castellanii* infection. Hum Pathol 12:573–576.

Bose K, Ghosh DK, Ghosh KN, Bhattacharya A, Das SR (1990) Characterization of potentially pathogenic free-living amoebae in sewage samples of Calcutta, India. Braz J Med Biol Res 23:1271–1278.

Bovee EC, Wilson DE, Telford SR Jr (1961) Some amoebas and ameboflagellates inquilinic in Florida reptiles. J Protozool 8:15.

Bowman EK, Vass AA, Mackowski R, Owen BA, Tyndall RL (1996) Quantitation of free-living amoebae and bacterial populations in eyewash stations relative to flushing frequency. Am Ind Hyg Assoc J 57:626–633.

Brasseur G, Favennec L, Perrine D, Chenu JP, Brasseur P (1994) Successful treatment of *Acanthamoeba* keratitis by hexamidine. Cornea 13:459–462.

Brown TJ, Cursons RTM (1977) Pathogenic free-living amebae (PFLA) from frozen swimming areas in Oslo, Norway. Scand J Infect Dis 9:237–240.

Brown TJ, Cursons RTM, Keys EA, Marks M, Miles M (1983) The occurrence and distribution of pathogenic free-living amoebae in thermal areas of the North Island of NZ N Z J Mar Freshw Res 17:59–69.

Butt CG (1966) Primary amebic meningoencephalitis. N Engl J Med 274:1473–1476.
Cerva L (1989) *Acanthamoeba culbertsoni* and *Naegleria fowleri*: occurrence of antibodies in man. J Hyg Epidemiol Microbiol Immunol 33:99–103.
Chang JCH, Ossoff SF, Lobe DC, Dorfman MH, Dumais CM, Qualls RG, Johnson JD (1985) UV inactivation of pathogenic and indicator microorganisms. Appl Environ Microbiol 49:1361–1365.
Cheng KH, Leung SL, Hoekman HW, Beekhuis WH, Mulder PGH, Geerards AJM (1999) Incidence of contact-lens-associated microbial keratitis and its related morbidity. Lancet 354:181–185.
Chu DM, Miles H, Toney D, Ngyuen C, Marciano-Cabral F (1998) Amebicidal activity of plant extracts from Southeast Asia on *Acanthamoeba* spp. Parasitol Res 84:746–752.
Cirillo JD, Falkow S, Tompkins LS, Bermudez LE (1997) Interaction of *Mycobacterium avium* with environmental amoebae enhances virulence. Infect Immun 65:3759–3767.
Cirillo JD, Cirillo SL, Yan L, Bermudez LE, Falkow S, Tompkins LS (1999) Intracellular growth in *Acanthamoeba castellanii* affects monocyte entry mechanisms and enhances virulence of *Legionella pneumophila*. Infect Immun 67:4427–4434.
Cohen EJ, Buchanan HW, Laughrea PA, Adams CP, Galentine PG, Visvesvara GS, Folberg R Arentsen JJ, Laibson PR (1985) Diagnosis and management of *Acanthamoeba* keratitis. Am J Ophthalmol 100:389–395.
Connor R, Hay J, Mead AJC, Seal DV (1993) Reversal of inhibitory effects of *Acanthamoeba castellanii* lysate for *Legionella pneumophila* using catalase. J Microbiol Methods 18:311–316.
Contact Lens Council (2000) <www.contactlenses.org>
Crespo EP, Mallen MM, Monica MP, Ares P, Fernandez MCA, Combarro MPC (1990) Isolation of amoebae of the genera *Naegleria* and *Acanthamoeba* from public fountains in Galicia (NW Spain). Water Air Soil Pollut 53:103–111.
Culbertson CG (1971) The pathogenicity of soil amebas. Annu Rev Microbiol 25:231–254.
Cursons RT, Brown TJ, Keys EA (1980a) Effect of disinfectants on pathogenic free-living amoebae: in axenic conditions. Appl Environ Microbiol 40:62–66.
Cursons RT, Brown TJ, Keys EA, Moriarty KM, Till D (1980b) Immunity to pathogenic free-living amoebae: role of humoral antibody. Infect Immun 29:401–407.
Daggett P-M (1982) In: Colwell RR (ed) Microbial Hazards of Diving in Polluted Waters: A Proceedings. University of Maryland Sea Grant Publ UM-SG-TS-82–01. University of Maryland, College Park, MD, pp 39–42.
D'Aversa G, Stern GA, Driebe WT Jr (1995) Diagnosis and successful medical treatment of *Acanthamoeba* keratitis. Arch Ophthalmol 113:1120–1123.
de Jonckheere J, Van de Voorde H (1976) Differences in destruction of cysts of pathogenic and nonpathogenic *Naegleria* and *Acanthamoeba* by chlorine. Appl Environ Microbiol 31:294–297.
de Jonckheere JF (1979a) Pathogenic free-living amoebae in swimming pools: survey in Belgium. Ann Microbiol (Inst Pasteur) 130B:205–212.
de Jonckheere JF (1979b) Occurrence of *Naegleria* and *Acanthamoeba* in aquaria. Appl Environ Microbiol 38:590–593.
de Jonckheere JF (1987) Taxonomy. In: Rondanelli EG (ed) Amphizoic Amoebae and Human Pathology. Piccin Nuova Libraria, Padua, Italy, pp. 25–48.
de Jonckheere JF, Michel R (1988) Species identification and virulence of *Acanthamoeba* strains from human nasal mucosa. Parasitol Res 74:314–316.

Detterline JL, Wilhelm WE (1991) Survey of pathogenic *Naegleria fowleri* and thermotolerant amebas in federal recreational waters. Trans Am Microsc Soc 110:244–261.

Di Gregorio C, Rivasi F, Mongiardo N, De Rienzo B, Wallace S, Visvesvara GS (1992) *Acanthamoeba* meningoencephalitis in a patient with acquired immunodeficiency syndrome. Arch Pathol Lab Med 116:1363–1365.

Dive D, Delattre JM, Leclerc H (1982) Occurrence of thermotolerant amoebae in an electric power plant cooling pond. J Therm Biol 7:11–14.

Dowling JN, Saba AK, Glew RH (1992) Virulence factors of the family *Legionellacae*. Microbiol Rev 56:32–60.

Dwivedi JN, Singh CM (1965) Pulmonary lesions in an Indian buffalo associated with *Acanthamoeba* sp. Indian J Microbiol 5:31–34.

Dyková I, Lom J, Macháčková B, Sawyer TK (1996) Amoebic infections in goldfishes and granulomatous lesions. Folia Parasitol 43:81–90.

el Sibae MM (1993) Detection of free-living amoeba (*Acanthamoeba polyphaga*) in the air conditioning systems. J Egypt Soc Parasitol 23:687–690.

EPA (U.S. Environmental Protection Agency) (1994) Enhanced surface water treatment rule. Fed Reg 59 FR 38832, July 29, 1994.

EPA (1998) Demographic distribution of sensitive population groups. Office of Science and Technology, Human Ecological Effects Division, EPA, Washington, DC.

EPA (2003) Health Effects Support Document for Acanthamoeba, EPA-822-R-03-012.

Ferrante A (1991) Free-living amoebae: pathogenicity and immunity. Parasite Immunol (Oxf) 13:31–47.

Ficker L (1988) *Acanthamoeba* keratitis—the quest for a better prognosis. Eye 2 (suppl): S37–S45.

Fields BS (1993) *Legionella* and protozoa: interaction of a pathogen and its natural host. In: Barbaree JM, Breiman RF, Dufour AP (eds) *Legionella* Current status and Emerging Perspectives. American Society for Microbiology, Washington, DC, p 129.

Fields BS, Sanders GN, Barbaree JM (1989) Intracellular multiplication of *Legionella pneumophila* in amoebae isolated from hospital hot water baths. Curr Microbiol 18: 131–137.

Franke ED, Mackiewicz S (1982) Isolation of *Acanthamoeba* and *Naegleria* from the intestinal contents of freshwater fishes and their potential pathogenicity. J Parasitol 68:164–166.

Friedland LR, Raphael SA, Deutsch ES, Johal J, Martyn LJ, Visvesvara GS, Lischner HW (1992) Disseminated *Acanthamoeaba* infection in a child with symptomatic human immunodeficiency virus infection. Pediatr Infect Dis J 11:404–407.

Fritsche TR, Sobek D, Gautom RK (1998) Enhancement of in vitro cytopathogenicity by *Acanthamoeba* spp. following acquisition of bacterial endosymbionts. FEMS Microbiol Lett 166:231–236.

Fritsche TR, Horn M, Seyedirashsti S, Gautom R, Schleifer K-H, Wagner M (1999) *In situ* detection of novel bacterial endosymbionts of *Acanthamoeba* spp. phylogenetically related to members of the order Rickettsiales. Appl Environ Microbiol 65:206–212.

Gardner HA, Martinez AJ, Visvesvara GS, Sotrel A (1991) Granulomatous amebic encephalitis in an AIDS patient. Neurology 41:1993–1995.

Gogate A, Singh BN, Deodhar LP, Jhala HI (1984) Primary amoebic meningo-encephalitis caused by *Acanthameoba*. J Postgrad Med 30:125–128.

Gonzalez MM, Gould E, Dickinson G, Martinez AJ, Visvesvara G, Cleary TJ, Hensley

GT (1986) Acquired immunodeficiency syndrome associated with *Acanthamoeba* infection and other opportunistic organisms. Arch Pathol Lab Med 110:749–751.

Gordon SM, Steinberg JP, DuPuis MH, Kozarsky PE, Nickerson JF, Visvesvara GS (1992) Culture isolation of *Acanthamoeba* species and leptomyxid amebas from patients with amebic meningoencephalitis, including two patients with AIDS. Clin Infect Dis 15:1024–1030.

Griffin JL (1972) Temperature tolerance of pathogenic and nonpathogenic free-living amoebas. Science 178:869–870.

Gullett J, Mills J, Hadley K, Podemski B, Pitts L, Gelber R (1979) Disseminated granulomatous *Acanthamoeba* infection presenting as an unusual skin lesion. Am J Med 67: 891–896.

Hamadto HH, Aufy SM, El-Hayawan IA, Saleh MH, Nagaty IM (1993) Study of free living amoebae in Egypt. J Egypt Soc Parasitol 23:631–637.

Hamburg A, De Jonckheere JF (1980) Amoebic keratitis. Ophthalmologica 181:74–80.

Hay J, Seal DV (1994a) Surveying for legionnaire's disease bacterium. Curr Opin Infect Dis 7:479–483.

Hay J, Seal DV (1994b) Monitoring of hospital water supplies for *Legionella*. J Hosp Infect 26:75–78.

Hay J, Seal DV, Billcliffe B, Freer JH (1995) Non-culturable *Legionella pneumophila* associated with *Acanthamoeba castellanii* detection of the bacterium using DNA amplification and hybridization. J Appl Bacteriol 78:61–65.

Hiti K, Walochnik J, Faschinger C, Haller-Schober EM, Aspock H (2001) Microwave treatment of contact lens cases contaminated with acanthamoeba. Cornea 20:467–470.

Hoffler AS, Rubel LR (1974) Free-living amoebae identified by cytologic examination of gastrointestinal washings. Acta Cytol 18:59–61.

Houang E, Lam D, Fan D, Seal D (2001) Microbial keratitis in Hong Kong: relationship to climate, environment and contact-lens disinfection. Trans R Soc Trop Med Hyg 95:361–367.

Illingworth CD, Cook SD (1998) *Acanthamoeba* keratitis. Surv Ophthalmol 42:493–508.

Illingworth CD, Cook CD, Karabatsas CH, Easty DL (1995) *Acanthamoeba* keratitis: risk factors and outcome. Br J Ophthalmol 79:1078–1082.

Ishibashi Y, Matsumoto Y, Kabata T, Watanabe R, Hommura S, Yasuraoka K, Ishii K (1990) Oral itraconazole and topical miconazole with débridement for *Acanthamoeba* keratitis. Am J Ophthalmol 109:121–126.

Jadin J-B, Willaert E, Hermanne J (1973) Présence d'amibes limax dans l'intestin de l'homme et des animaux Académie Royale des Sciences d'Outre Mer. Bull Seances (Bruxelles) 1973(3):520–526.

Janitschke K, Werner H, Müller G (1980) Examinations on the occurrence of free-living amoebae with possible pathogenic traits in swimming pools. Zentbl Bakteriol Abt 1 Orig B 170:108–122.

Janitschke K, Martinez AJ, Visvesvara GS, Schuster F (1996) Animal model *Balamuthia mandrillaris* CNS infection: contrast and comparison in immunodeficient and immunocompetent mice: a murine model of "granulomatous" amebic encephalitis. J Neuropathol Exp Neurol 55:815–821.

John DT, Howard MJ (1995) Seasonal distribution of pathogenic free-living amebae in Oklahoma waters. Parasitol Res 81:193–201.

Johns KJ, O'Day DM, Feman SS (1988) Chorioretinitis in the contralateral eye of a patient with *Acanthamoeba* keratitis. Ophthalmology 95:635–639.

Jones BR, McGill JI, Steele ADM (1973) Recurrent suppurative kerato-uveitis with loss of eye due to infection by *Acanthamoeba castellanii*. Trans Ophthalmol Soc UK 95: 210–213.

Jones DB, Visvesvara GS, Robinson NM (1975) *Acanthamoeba polyphaga* keratitis and *Acanthamoeba* uveitits associated with fatal menengoencephalitis. Trans Ophthalmol Soc UK 95:221–232.

Kadlec V (1978) The occurrence of amphizoic amebae in domestic animals. J Protozool 25:235–237.

Khan NA (2001) Pathogenicity, morphology, and differentiation of *Acanthamoeba*. Curr Microbiol 43:391–395.

Khan NA, Jarroll EL, Paget TA (2001) *Acanthamoeba* can be differentiated by the polymerase chain reaction and simple plating assays. Curr Microbiol 43:204–208.

Kilvington S (1989) Moist-heat disinfection of pathogenic *Acanthamoeba* cysts. Lett Appl Microbiol 9:187–189.

Kilvington S, Price J (1990) Survival of *Legionella pneumophila* within cysts of *Acanthamoeba polyphaga* following chlorine exposure. J Appl Bacteriol 68:519–525.

Kilvington S, Larkin DFP, White DG, Beeching JR (1990) Laboratory investigation of *Acanthamoeba* keratitis. J Clin Microbiol 28:2722–2725.

King CH, Shotts EB Jr, Wooley RE, Porter KG (1988) Survival of coliforms and bacterial pathogens within protozoa during chlorination. Appl Environ Microbiol 54:3023–3033.

Kingston D, Warhurst DC (1969) Isolation of ameba from air. J Med Microbiol 2:27–36.

Kyle DE, Noblet GP (1987) Seasonal distribution of thermotolerant free-living amoebae. II. Lake Issaqueena. J Protozool 34:10–15.

Larkin DF, Kilvington S, Dart JK (1992) Treatment of *Acanthamoeba* keratitis with polyhexamethylene biguanide. Ophthalmology 99:185–191.

Lawande RV, Abraham SN, John I, Egler LJ (1979) Recovery of soil amebas from the nasal passages of children during the dusty harmattan period in Zaria. Am J Clin Pathol 71:201–203.

Ledee DR, Hay J, Byers TJ, Seal DV, Kirkness CM (1996) *Acanthamoeba griffini*: molecular characterization of new corneal pathogen. Invest Ophthalmol Vis Sci 37:544–550.

Lengy J, Jakovljevich R, Talis B (1971) Recovery of a hartmannelloid amoeba from a purulent ear discharge. Harefuah 80:23–24.

Ly TM, Muller HE (1990) Ingested *Listeria monocytogenes* survive and multiply in protozoa. J Med Microbiol 33:51–54.

Lyons TB, Kapur R (1977) Limax amoebae in public swimming pools of Albany, Schenectady, and Rensselaer counties, New York: their concentration, correlations, and significance. Appl Environ Microbiol 33:551–555.

Ma P, Visvesvara GS, Martinez AJ, Theodore FH, Daggett P-M, Sawyer TK (1990) *Naegleria* and *Acanthamoeba* infections: review. Rev Infect Dis 12:490–513.

Mannis MJ, Tamaru R, Roth AM, Burns M, Thirkill C (1986) *Acanthamoeba* sclerokeratitis: determining diagnostic criteria. Arch Ophthalmol 104:1313–1317.

Marciano-Cabral F, Puffenbarger R, Cabral GA (2000) The increasing importance of *Acanthamoeba* infections. J Eukaryot Microbiol 47:29–36.

Martinez AJ (1977) Free-living amebic meningoencephalitides: comparative study. Neurol Psihiatr Neurochir (Buchar) 18:391–401.

Martinez AJ (1980 Is *Acanthamoeba* encephalitis an opportunistic infection? Neurology 30:567–574.

Martinez AJ (1982) Acanthamoebiasis and immunosuppression. Case report. J Neuropathol Exp Neurol 41:548–557.
Martinez AJ (1985) Free-Living Amebas: Natural History, Prevention, Diagnosis, Pathology, and Treatment of Disease. CRC Press, Boca Raton, FL.
Martinez AJ (1993) Free-living amebas: infection of the central nervous system. Mt Sinai J Med 60:271–278.
Martinez AJ, Janitschke K (1985) *Acanthamoeba*, an opportunistic microorganism: a review. Infection 13:251–256.
Martinez AJ, Visvesvara GS (1991) Laboratory diagnosis of pathogenic free-living amoebas: *Naegleria, Acanthamoeba*, and leptomyxid. Clin Lab Med 11:861–872.
Martinez AJ, Visvesvara GS (1997) Free-living, amphizoic and opportunistic amebas. Brain Pathol 7:583–598.
Martinez AJ, Visvesvara GS (2001) *Balamuthia mandrillis* infection. J Med Microbiol 50:205–207.
Martinez AJ, Sotelo-Avila C, Garcia-Tamayo J, Moron JT, Willaert E, Stamm WP (1977) Meningoencephalitis due to *Acanthamoeba* sp.: pathogenesis and clinicopathological study. Acta Neuropathol 37:183–191.
Martinez AJ, Visvesvara GS, Chandler FW (1997) Free-living amebic infections. In: Connor DH, Chandler FC, Schwartz DA, Manz HJ, Lack EE (eds) Pathology of Infectious Diseases, vol ll. Appleton & Lange, Norwalk, pp. 1163–1176.
Mathers WD, Sutphin JE, Folberg R, Meier PA, Wenzel RP, Elgin RG (1996) Outbreak of keratitis presumed to be caused by *Acanthamoeba*. Am J Ophthalmol 121:129–142.
McConnell EE, Garner FM, Kirk JH (1968) Hartmannellosis in a bull. Pathol Vet 5:1–6.
Mehta AP, Guirges SY (1979) Acute amoebic dysentery due to free-living amoebae treated with metronidazole. J Trop Med Hyg 82:134–136.
Meier PA, Mathers WD, Sutphin JE, Folberg R, Hwang T, Wenzel RP (1998) An epidemic of presumed *Acanthamoeba* keratitis that followed regional flooding. Results of a case-control investigation. Arch Ophthalmol 116:1090–1094.
Michel R, Müller KD, Amann R, Schmid EN (1998) *Legionella*-like slender rods multiplying within a strain of *Acanthamoeba* sp isolated from drinking water. Parasitol Res 60:84–88.
MMWR (1987) *Acanthamoeba* keratitis in soft-contact-lens wearers. MMWR (Morb Mortal Wkly Rep 36:397–404. Centers for Disease Control, Atlanta, GA.
Moore MB, McCulley JP (1989) *Acanthamoeba* keratitis associated with contact lenses: six consecutive cases of successful management. Br J Opthalmol 73:272–275.
Moore MB, McCulley JP, Luckenbach M, Gelender H, Newton C, McDonald MB, Visvesvara GS (1985) *Acanthamoeba* keratitis associated with soft contact lenses. Am J Ophthalmol 100:396–403.
Moura H, Wallace S, Visvesvara GS (1992) *Acanthamoeba healyi*, n.sp., and isosenzyme and immunoblot profiles of *Acanthamoeba* spp., groups II and III. J Protozool 39: 573–583.
Munson DM (1993) The distribution of cyst-forming amoebae from inshore and freshwater Bermuda sediments. Trans Am Microsc Soc 112:88.
Nagington J, Watson PG, Playfair TJ, McGill J, Jones BR, Steele ADM (1974) Amoebic infection of the eye. Lancet 2:1537–1540.
Nerad T (1993) Catalogue of Protists, 18th Ed. American Type Culture Collection, Manassas, VA.

O'Dell WD (1979) Isolation, enumeration and identification of amebae from a Nebraska lake. J Protozool 26:265-269.
Ofori-Kwakye SK, Sidebottom DG, Fischer HJ, Visvesvara GS (1986) Granulomatous brain tumor caused by *Acanthamoeba*: a case report. J Neuosurg 64:505-509.
O'Malley ML, Lear DW, Adams WN, Gaines J, Sawyer TK, Lewis EJ (1982) Microbial contamination of continental shelf sediments by wastewater. J Water Pollut Control Fed 54:1311-1317.
Ormerod LD, Smith RE (1986) Contact lens-associated microbial keratitis. Arch Ophthalmol 104:79-83.
Page FC (1967) Re-definition of the genus *Acanthamoeba* with description of three species. J Protozool 14:709-724.
Page FC (1976) An Illustrated Key to Freshwater and Soil Amoebae. Freshwater Biological Association, Ambleside, Cumbria.
Page FC (1988) A New Key to Freshwater and Soil Gymnamoebae with Instructions for Culture. Freshwater Biological Association, Ambleside, Cumbria.
Park CH, Iyengar V, Hefter L, Pestaner JP, Vandel NM (1994) Cutaneous *Acanthamoeba* infection associated with acquired immunodeficiency syndrome. Lab Med 25:386-388.
Penas-Ares M, Paniagua-Crespo E, Madrinan-Choren R, Marti-Mallen M, Arias-Fernandez MC (1994) Isolation of free-living pathogenic amoebae from thermal spas in NW Spain. Water Air Soil Pollut 78:83-90.
Pussard M, Pons R (1977) Morphologie de la paroi kystique et taxonomie du genre *Acanthamoeba* (Protozoa, Amoebida). Protistologica 13:557-598.
Radford CF, Lehmann OJ, Dart JKG (1998) *Acanthamoeba* keratitis: multicentre survey in England 1992-6. Br J Ophthalmol 82:1387-1392.
Regli S, Rose JB, Haas CN, Gerba CP (1991) Modeling the risk from *Giardia* and viruses in drinking water. J Am Water Works Assoc 83:76-84.
Rivera F, Ortega A, Lopez-Ochotorena E, Paz ME (1979) A quantitative morphological and ecological study of protozoa polluting tap water in Mexico City. Trans Am Microsc Soc 98:465-469.
Rivera F, Galván M, Robles E, Leal P, González L, Lacy AM (1981) Bottled mineral waters polluted by protozoa in Mexico. J Protozool 28:54-56.
Rivera F, Ramírez P, Vilaclara G, Robles E, Medina F (1983) A survey of pathogenic and free-living amoebae inhabiting swimming pool water in Mexico City. Environ Res 32:205-211.
Rivera F, Roy-Ocotla G, Rosas I, Ramirez E, Bonilla P, Lares F (1987) Amoebae isolated from the atmosphere of Mexico City and environs. Environ Res 42:149-154.
Rivera F, Lares F, Ramirez E, Bonilla P, Rodriquez S, Labastida A, Ortiz R, Hernandez D (1991) Pathogenic *Acanthamoeba* isolated during an atmospheric survey in Mexico City. Microbiology of *Acanthamoeba* (extended abstract). Rev Infect Dis 13(suppl 5): S388-S389.
Rodriguez-Zaragoza S (1994) Ecology of free-living amoebae. Crit Rev Microbiol 20: 225-241.
Rohr U, Weber S, Michel R, Selenka F, Wilhelm M (1998) Comparison of free-living amoebae in hot water systems of hospitals with isolates from moist sanitary areas by identifying genera and determining temperature tolerance. Appl Environ Microbiol 64:1822-1824.
Rowan-Kelly B, Ferrante A (1984) Immunization with killed *Acanthamoeba culbertsoni*

antigen and amoeba culture supernatant antigen in experimental *Acanthamoeba* meningoencephalitis. Trans R Soc Trop Med Hyg 78:179–182.
Rowbotham TJ (1980) Preliminary report on the pathogenicity of *Legionella pneumophila* for freshwater and soil amoebae. J Clin Pathol 33:1179–1183.
Rude RA, Jackson GJ, Bier JW, Sawyer TK, Risty NG (1984) Survey of fresh vegetables for nematodes, amoebae, and *Salmonella*. J Assoc Offic Anal Chem 67:613–615.
Sawyer TK (1992) Distribution of microbial agents in marine ecosystems as a consequence of sewage-disposal practices. In: Rosenfield A, Mann R (eds) Disposal of Living Organisms into Aquatic Ecosystems. Maryland Sea Grant College, University of Maryland Systems, College Park, MD, pp. 239–262.
Sawyer TK, Lewis EJ, Galassa M, Lear DW, O'Malley ML, Adams WN, Gaines J (1982) Pathogenic amoebae in ocean sediments near wastewater sludge disposal sites. J Water Pollut Control Fed 54:1318–1323.
Sawyer TK, Nerad TA, Visvesvara GS (1992) *Acanthamoeba jacobsi* sp. n. (Protozoa: Acanthamoebidae) from sewage-contaminated ocean sediments. Proc Helminthol Soc Wash 59:223–226.
Schaumberg DA, Snow KK, Dana MR (1998) The epidemic of *Acanthamoeba* keratitis: where do we stand? Cornea 17:3–10.
Schuster FL, Jacob LS (1992) Effects of magainins on ameba and cyst stages of *Acanthamoeba polyphaga*. Antimicrob Agents Chemother 36:1263–1271.
Schuster FL, Visvesvara GS (1998) Efficacy of novel antimicrobials against clinical isolates of opportunistic amebas. J Eukaryot Microbiol 45:612–618.
Seal D, Stapleton F, Dart J (1992) Possible environmental sources of *Acanthamoeba* spp. in contact lens wearers. Br J Ophthalmol 76:424–427.
Seal D, Hay J, Kirkness C, Morrell A, Booth A, Tullo A, Ridgway A, Armstrong M (1996) Successful medical therapy of *Acanthamoeba* keratitis with topical chlorhexidine and propamidine. Eye 10:413–421.
Seal DV (2000) Contact-lens-associated microbial keratitis in the Netherlands and Scotland. Lancet 355:143–144.
Simitzis-LeFlohic AM, Chastel C (1982) Le petits mammiferes sauvages: vecteurs d'amibes libres? Med Trop 42:275–279.
Singh BN (1952) Nuclear division in nine species of small free-living amoebae and its bearing on the classification of the order Amoebida. Philos Trans R Soc Lond 236:405–460.
Singh BN, Das SR (1972) Occurrence of pathogenic *Naegleria aerobia, Hartmannella culbertsoni* and *H. rhysodes* in sewage sludge samples of Lucknow. Curr Sci 41:277–281.
Singh U, Petri WA (2000) Free-living amebas. In: Mandell GL, Bennett JE, Dolin R (eds) Principle and Practice of Infectious Disease, 5th Ed. Churchill Livingstone, Philadelphia, pp 2811–2817.
Stehr-Greene JK, Bailey TM, Brandt FH, Carr JH, Bond WW, Visvesvara GS (1987) *Acanthamoeba* keratitis in soft contact lens wearers: a case-control study. JAMA 258:57–60.
Stothard DR, Schroeder-Diedrich JM, Awwad MH, Gast RJ, Ledee DR, Rodriguez-Zaragoza S, Dean CL, Fuers PA, Byers TJ (1998) The evolutionary history of the genus *Acanthamoeba* and the identification of eight new 18S rRNA gene sequence types. J Eukaryot Microbiol 45:45–54.
Taylor PW (1977) Isolation and experimental infection of free-living amebae in freshwater fishes. J Parasitol 63:232–237.

Teknos TN, Poulin MD, Laruentano AM, Li KK (2000) *Acanthamoeba* rhinosinusitis: characterization, diagnosis, and treatment. Am J Rhinol 14:387–391.

Telang GH, Scola F, Kantor GR, Stieritz D, Reboli AC (1996) Disseminated *Acanthamoeba* infection in a patient with AIDS. Infect Dis Clin Pract 5:387–389.

Thamprasert K, Khunamornpong S, Morakote N (1993) *Acanthamoeba* infection of peptic ulcer. Ann Trop Med Parasitol 87:403–405.

Theodore FH, Jakobiec FA, Juechter KB, Ma P, Troutman RC, Pang PM, Iwamoto T (1985) The diagnostic value of a ring infiltrate in *Acanthamoeba* keratitis. Ophthalmology 92:1471–1479.

Tyndall RL, Lyle MM, Ironside KS (1987) The presence of free-living amoebae in portable and stationary eye wash stations. Am Ind Hyg Assoc J 48:933–934.

van Klink F, Alizadeh H, Stewart GL, Pidherney MS, Silvany RE, YuGuang H, McCulley JP, Niederkorn JY (1992) Characterization and pathogenic potential of a soil isolate and an ocular isolate of *Acanthamoeba castellanii* in relation to *Acanthamoeba* keratitis. Curr Eye Res 11:1207–1220.

Visvesvara GS, Jones DB, Robinson NM (1975) Isolation, identification, and biological characterization of *Acanthamoeba polyphaga* from a human eye. Am J Trop Med Hyg 24:784–790.

Visvesvara GS, Mirra SS, Brandt FH, Moss DM, Mathews HM, Martinez AJ (1983) Isolation of two strains of *Acanthamoeba castellanii* from human tissue and their pathogenicity and isoenzyme profiles. J Clin Microbiol 18:1405–1412.

Visvesvara GS, Neafie RC, Martinez AJ (1997) Pathogenic and opportunistic free-living amoebae. In: Horsburgh CR Jr, Nelson AM (eds) Pathology of Emerging Infections. American Society of Microbiology, Washington, DC, pp 257–267.

Volker-Dieben HJM, Bos HJ, Kok-van Alphen CC (1980) Amebic keratitis in a windsurfer. Ned Tijdschr Geneeskd 124:1147–1151.

Walker PL, Prociv P, Gardiner WG, Moorehead DE (1986) Isolation of free-living amoebae from air samples and an air conditioner filter in Brisbane. Med J Aust 145:175.

Walochnik J, Haller-Schober E, Kolli H, Picher O, Obwaller A, Aspock H (2000) Discrimination between clinically relevant and nonrelevant *Acanthamoeba* strains isolated from contact lens-wearing keratitis patients in Austria. J Clin Microbiol 38: 3932–3936.

Wiley CA, Safrin RE, Davis CE, Lampert PW, Braude AI, Martinez AJ, Visvesvara GS (1987) *Acanthamoeba* meningoencephalitis in a patient with AIDS. J Infect Dis 155: 130–133.

Wilhelmus KR, Jones DB (1991) Program planning for research on *Acanthamoeba*. Rev Infect Dis Suppl 5:S446–S450.

Wilson DE, Bovee EC, Bovee GJ, Telford SR Jr (1967) Induction of amebiasis in tissues of white mice and rats by subcutaneous inoculation of small free-living, inquilinic, and parasitic amebas with associated coliform bacteria. Exp Parasitol 21:277–286.

Wright P, Warhurst D, Jones BR (1985) *Acanthamoeba* keratitis successfully treated medically. Br J Ophthalmol 69:778–782.

Yamaura H, Shirasaka R, Matsumoto K, Kuwabara M, Tsuji M, Horikami H (1993) Isolation of *Acanthamoeba* from sandboxes in Tokyo and Hiroshima City. Jpn J Parasitol 42:361–364.

Yang YF, Matheson M, Dart JK, Cree IA (2001) Persistence of acanthamoeba antigen following acanthamoeba keratitis. Br J Ophthalmol 85:277–280

Manuscript received October 30, 2002; accepted February 12, 2003.

Arsenic Hazards to Humans, Plants, and Animals from Gold Mining

Ronald Eisler

Contents

I. Introduction	133
II. Arsenic Sources to the Biosphere from Gold Mining	133
III. Arsenic Risks to Human Health	135
IV. Arsenic Concentrations in Abiotic Materials and Biota near Gold Extraction Facilities	137
V. Arsenic Effects on Sensitive Species	142
VI. Proposed Arsenic Criteria	155
Summary	159
Acknowledgments	160
References	160

I. Introduction

Arsenic contamination of the biosphere from various gold mining and refining operations may jeopardize the health and well-being of biological communities. This review documents the sources and extent of arsenic discharges to the environment associated with gold mining operations; arsenic risks to human health, with emphasis on gold miners, gold refinery workers, and children residing near gold mining and refining activities; arsenic concentrations in biota and abiotic materials near gold extraction and refining facilities; lethal and sublethal effects of different chemical forms of arsenic to representative species of flora and fauna; and proposed arsenic criteria for the protection of human health and selected natural resources. It is part of a larger work in progress on gold ecotoxicology.

II. Arsenic Sources to the Biosphere from Gold Mining

Gold-bearing ores worldwide contain variable quantities of sulfide and arsenic compounds that interfere with efficient gold extraction using current cyanidation technology. Arsenic occurs in many types of Canadian gold ore deposits, mainly as arsenopyrite (FeAsS), niccolite (NiAs), cobaltite (CoAsS), tennantite

---

Communicated by George W. Ware.

R. Eisler
U.S. Geological Survey, Patuxent Wildlife Research Center, 11510 American Holly Drive, Laurel, MD 20708-4019, U.S.A.

[(Cu,Fe)$_{12}$As$_4$S$_{13}$], enargite (Cu$_3$AsS$_4$), orpiment (As$_2$S$_3$), and realgar (AsS) (Azcue et al. 1994). Some gold-containing ores in Columbia, South America, contain up to 32% of arsenic-bearing minerals, and surrounding sediments may hold as much as 6300 mg As/kg DW (Grosser et al. 1994).

Arsenic enters the environment from a variety of sources associated with gold mining, including waste soil and rocks, tailings, atmospheric emissions from ore roasting, and bacterially enhanced leaching. The combination of opencast mining and heap leaching generates large quantities of waste soil and rock (overburden) and residual water from ore concentrations (tailings). The wastes, especially the tailings, are rich sources of arsenic (Greer 1993). In Nova Scotia, for example, about 3 million tons of tailings, containing 20,700 kg arsenic, were left from gold mining activities between 1860 and 1945. Tailings tend to diffuse into the surrounding environment over time, with subsequent spread of arsenic contamination (Wong et al. 1999).

Discharges from gold mines into the Humboldt Sink, Nevada, sometimes exceed water quality regulations mandated for arsenic [U.S. Bureau of Land Management (USBLM) 2000]. In the Black Hills of South Dakota, a cluster of 11 abandoned gold mines discharged up to 10,000 kg arsenopyrites daily into nearby creeks (Rahn et al. 1996). The present treatment of gold mine tailings to reduce arsenic availability to the environment involves peroxide addition to oxidize cyanide to cyanate, ferric sulfate and lime addition to precipitate arsenic as ferric arsenate (FeAsO$_4$), and polyacrylamide flocculent addition to enhance sedimentation (Bright et al. 1994, 1996).

As discussed later, roasting of some types of gold-containing ores to remove sulfur has resulted in significant atmospheric emissions of arsenic trioxide (As$_2$O$_3$) and sulfur oxides (Ripley et al. 1996). Arsenic was at one time extracted as a by-product in many gold mines and sold mainly for the manufacture of pesticides; however, this use is no longer profitable (Azcue et al. 1994). In Fairbanks, Alaska, some groundwaters are contaminated with arsenic from gold mining activities 30 yr earlier and considered unsafe for drinking; bacteria associated with arsenic in mine drainage may accelerate the rate at which arsenic leaches from the sediment into groundwater (Pain 1987).

Refractory gold ores are those that are not free milling and require pretreatment before cyanide leaching (Adams et al. 1999). In most refractory ores, gold is locked in sulfides or is substituted in the sulfide mineral lattice. Commercial treatment of these ores involves roasting to destroy the sulfide minerals and liberate the gold, the calcine being treated by conventional cyanidation. In the treatment of ores containing arsenopyrite, environmental contamination may occur from release of sulfur dioxide and arsenic trioxide:

$$2FeAsS + 5O_2 \rightarrow 2SO_2 + Fe_2O_3 + As_2O_3$$

In Canada, roasting has been largely discontinued; however, at least three operating facilities in that country were still using this practice in 1992 (Ripley et al. 1996). In Ghana, arsenic trioxides and other arsenic oxides from roasting of gold ores that were lost to the atmosphere were subsequently deposited in rainfall, causing extensive arsenic contamination of soil, vegetation, crops, humans, rivers, and livestock (Golow et al. 1996). Despite the pollution aspects, roasting is still recommended as the most cost-effective method for the treatment of refractory gold ores (Adams et al. 1999). To reduce arsenic emissions, new processes have been developed for the treatment of refractory ores, including pressure oxidation, biooxidation, whole ore roasting, ultrafine grinding, nitric acid oxidation, and fine milling combined with low-pressure oxidation. In whole ore roasting, pressure oxidation, and biooxidation, arsenic is fixed as basic ferric arsenate instead of As_2O_3 (Adams et al. 1999). Other operations have extracted the arsenic through flotation, cycloning, alkaline chlorination, ferric ion precipitation, bioleaching and bacterial oxidation, and pressure oxidation using an autoclave (Ripley et al. 1996).

Bacterial decomposition of arsenopyrite assists in opening the molecular mineral structure, allowing access to the gold by cyanide. Arsenic can become a limiting factor in the bioleaching of arsenopyrite for the recovery of gold at high temperatures because of the formation of soluble As^{3+} and As^{5+} and their toxicity, especially that of As^{3+}, to strains of bacteria that were not resistant to arsenic (Hallberg et al. 1996). Biooxidation of difficult to treat gold-bearing arsenopyrite ores is now done commercially in aerated, stirred tanks with rapidly growing, arsenic-resistant bacterial strains of *Thiobacillus* spp., *Sulfolobus* sp., and *Leptospirullium* sp. (Ngubane and Baecker 1990; Agate 1996; Rawlings 1998). These obligate chemoautolithotrophic strains of bacteria obtain their energy through the oxidation of ferrous to ferric iron or through the reduction of inorganic sulfur compounds to sulfate.

Arsenic is often found as a mineral in combination with iron and sulfur. Oxidation of these insoluble forms results in the formation of arsenite (As^{3+}). In environments such as acid mine drainage of abandoned gold mines, As^{3+} concentrations ranged from 2 to 13 mg/L (Santini et al. 2000). The As^{3+} can then be oxidized to arsenate (As^{5+}). Both these soluble forms of arsenic are toxic to living organisms, especially inorganic arsenite. The chemical oxidation of arsenite to arsenate is slow compared to microbiological processes (Santini et al. 2000). Some species of bacteria protect against arsenic by reducing As^{5+} that has entered the cell to As^{3+} and then transporting As^{3+} out of the cell; however, arsenate reduction does not seem to support growth.

III. Arsenic Risks to Human Health

Beneficial uses of arsenic compounds in medicine have been known for at least 2400 yr. Inorganic arsenicals have been used for centuries, and organoarsenicals for at least a century, in the treatment of syphilis, yaws, amoebic dysentery, asthma, tuberculosis, leprosy, dermatoses, and trypanosomiasis (Asperger and Ceina-Cizmek 1999; Eisler 2000). The advent of penicillin and other newer drugs nearly eliminated the use of organic arsenicals as human therapeutic agents, although arsenical drugs are still used in treating African sleeping sickness and amoebic dysentery and in veterinary medicine to treat filariasis in dogs and blackhead in poultry (Eisler 2000).

By contrast, arsenic contamination of the environment, even at low levels of exposure, has potential human health hazards, including skin cancer, stomach cancer, respiratory tract cancer, hearing and vision impairment, melanosis, leukomelanosis, keratosis, hyperkeratosis, edema, gangrene, and extensive liver damage (Kabir and Bilgi 1993; Kusiak et al. 1993; Simonato et al. 1994; Huang and Dasgupta 1999; Matschullat et al. 2000). Arsenic-contaminated drinking water is a major health problem in Bangladesh and other parts of the Indian subcontinent as a result of arsenic-bearing sediments in contact with the aquifer. Ironically, the use of groundwater for drinking water was implemented to eliminate waterborne pathogens; this effort was initiated by international organizations led by the United Nations (Huang and Dasgupta 1999; Eisler 2000).

Canadian gold miners had an excess of mortality from carcinoma of the stomach and respiratory tract compared to other miners. The increased frequency of stomach cancer appeared 5–19 yr after they began gold mining in Ontario (Kusiak et al. 1993). A number of explanations are offered to account for the high death rate, including exposure to arsenic (Kusiak et al. 1993). Gold miners in Ontario with 5 yr or more gold mining experience before 1945 had a significantly increased risk of primary cancer of the trachea, bronchus, and lung (Kabir and Bilgi 1993). A minimum of 15 yr latency was recorded between first employment and diagnosis of lung cancer. Underground miners were exposed to air concentrations of 2.4–5.6 μg As/m^3 and had significantly elevated concentrations of arsenic in urine. For purposes of work-relatedness, it was concluded that arsenic exposure was one of several causes of primary lung cancer in the Ontario gold miners (Kabir and Bilgi 1993).

In France, a high incidence of neoplasms of the respiratory system among gold extraction and refinery workers was first reported in 1977 and again in 1985, and this appears related to occupational exposure (Simonato et al. 1994). Mine and smelter workers at this site were twice as likely to die of lung cancer than the general population. The lung cancer excess was strongly associated with exposure to soluble and insoluble forms of arsenic (Simonato et al. 1994). In Zimbabwe, arsenic exposure was implicated in lung cancer increase of gold miners (Boffetta et al. 1994).

Active gold mining in the state of Minas Gerais, Brazil, has been documented since the early 1700s (Matschullat et al. 2000). Three major gold deposits can be discerned within the volcanic sedimentary sequence of the Nova Lima group near the city of Belo Horizonte. In the 1990s, yearly gold production was around 6 t extracted from about 1 million t of ore. Most of the ores contained arsenopyrites with high potential for arsenic contamination. Although arsenic emissions from ore processing should be minimal because of modern control facilities, this was not the case here because of the overall poverty in the area. In addition, the local population used surface waters not only for fishing and gardening but frequently as their drinking water. Sources of arsenic to the biosphere included weathering of mine wastes via erosion, dissolution of arsenic-contaminated soils and tailings into surface waters and sediments, and smelting activities that released arsenic into the air through oxidation of arsenopyrites. In April 1998, 126 school children aged on average 9.8 years (8.7–10.9) yr in this southeastern Brazilian mining district had low urinary levels of cadmium (mean, 0.13; range, 0.04–0.35 µg/L), partly elevated concentrations of mercury (mean, 1.1; range, 0.1–16.5 µg/L), and generally elevated to high concentrations of arsenic (mean, 25.7; range, 2.2–106.0 µg/L). Of the total population, 20% showed elevated arsenic concentrations associated with future adverse health effects. Arsenic concentrations were high in local surface waters, soils, sediments, and mine tailings (Table 1), with arsenic-contaminated drinking water as the probable causative factor of elevated arsenic in urine (Matschullat et al. 2000).

Residents of La Oraya, Peru, experienced respiratory problems caused by arsenic and sulfur dioxide emissions released from an area smelter that processed gold and other ores; a soil sample collected 4 km downwind of the smelter contained 12,600 mg/kg surface arsenic as well as 22,000 mg/kg lead and 305 mg/kg cadmium (Da Rosa and Lyon 1997).

IV. Arsenic Concentrations in Abiotic Materials and Biota near Gold Extraction Facilities

Arsenic is a relatively common element that occurs in air, water, soil, and all living tissues (Eisler 2000). It ranks 20th in abundance in the Earth's crust, 14th in seawater, and 12th in the human body. Arsenic is a teratogen and carcinogen that can traverse placental barriers and produce fetal death and malformations in many species of mammals. Arsenic is carcinogenic in humans, but evidence for arsenic-induced carcinogenicity in other mammals is scarce. Arsenic concentrations are usually low [<1.0 mg/kg fresh weight (FW)] in most living organisms, but they are frequently elevated in marine biota, in which arsenic occurs as arsenobetaine and poses little risk to organisms or their consumers, and in plants and animals from areas that are naturally arseniferous or near anthropogenic sources (Eisler 2000).

Arsenic concentrations in samples collected near gold mining and processing facilities worldwide were elevated in sediments, sediment pore waters, water column, mine tailings, mine tailing drainage waters, soils, terrestrial plants (including edible plants used in human diets), aquatic plants, aquatic bivalve molluscs, terrestrial and aquatic insects, fishes, bird tissues, and human urine (see Table 1). Inorganic arsenicals are considered more toxic than organic arsenicals and trivalent arsenite (As^{3+}) compounds are more toxic than pentavalent arsenate (As^{5+}) compounds. Total arsenic, As^{3+}, and As^{5+} can now be measured under field conditions at a detection limit of 1 µg/L with a portable stripping voltammetric instrument using a gold film electrode (Huang and Dasgupta 1999).

Gold mining has been a major activity in Canada for more than a century (Azcue et al. 1994). Since 1921, Canada has ranked among the top three gold-producing nations. Abandoned gold mine tailings and waste rock contain large quantities of arsenic with high potential for adverse environmental effects. In

Table 1. Arsenic concentrations in biota and abiotic materials collected near gold mining and processing facilities.

Location, sample, and other variables	Concentration (mg total arsenic/ kg DW or FW)[a]	Reference[b]
South America		
Brazil: April 1998; southeastern gold mining districts		
Schoolchildren, age 8–11 yr; urine	0.026 (0.002–0.106) FW	1
Surface waters	0.031 (0.004–0.35) FW	1
Soils	200–800 DW	1
Sediments	350 (22–3,200) DW	1
Tailings	10,500 (300–21,000) DW	1
Columbia, stream sediments	Max. 6,300 DW	2
Ecuador: 1988; dry season, downstream of cyanide-gold mining area		
Water: measured vs. recommended	0.002–0.264 FW vs. <0.19 FW	2
Sediments: measured vs. recommended	403–7,700 DW vs. <17 DW	3
Peru: surface soils 4 km downwind of gold smelter	12,600 DW	4
North America		
British Columbia, Canada: site of underground gold mine; 1933–1964 (northeast shore of Jack of Clubs Lake)		
Tailings	>2,000 DW	5
Lake sediments	Max. 1,104 DW	5
Lake water	Max. 0.56 FW	5
Nova Scotia, Canada: stream waters at Goldenville mine; upstream vs. at mine discharge	0.03–0.05 FW vs. 0.23–0.25 FW	6
Yellowknife, NWT, Canada: 1990–1991; subarctic lakes; watershed contaminated with arsenic from effluent of two gold mines over several decades		
Surface sediments (gold content maximum 6.75 mg/kg DW)	2,186 (22–3,090) DW	7,8
Sediment pore waters	Max. 5.2 FW	8
Overlying water column	Max. 0.53–0.55 FW	7,8
United States		
Whitewood Creek, South Dakota (recipient of gold mine tailings 1876–1977) vs. reference site; 1987		
Sediments	764 DW vs. 18 DW	9
Aquatic insects, 4 species	73, 77, 278, and 625 DW vs. 1–16 DW	9

Table 1. (Continued).

Location, sample, and other variables	Concentration (mg total arsenic/ kg DW or FW)[a]	Reference[b]
Whitewood Creek (arsenic impacted from gold tailings containing an estimated 270,000 t arsenic between 1920 and 1977) vs. reference site in Casper, Wyoming		
Sediments, 1989	1,920 DW vs. 9 DW	10
House wren,		
Troglodytes aedon; 1997		
Eggs	<0.5 DW vs. <0.5 DW	10
Chicks		
Livers	2.9 (1.8–5.6) DW vs <0.5 DW	10
Diet (benthic insects)	103.0 DW vs. <0.5 DW	10
Africa		
Ghana		
Near gold ore processing facility vs. reference sites; topsoil		
Total arsenic	50 DW vs. 3–10 DW	11
As^{5+}	35 DW vs. no data	11
As^{3+}	15 DW vs. 1–2 DW	11
Near gold ore-roasting facility (17 t arsenic discharged to atmosphere/d) vs. reference site		
Cooked foods, edible portions		
Cassava, *Manihot esculenta*	2.7 DW vs. 1.9 DW	12
Plantain, *Musa paradisiaca*	3.4 DW vs. 3.0 DW	12
Other cooked foods	2.4 DW vs. 1.4 DW	12
Oil palm fruit, *Elaeis guineensis*	Max. 5.9 DW vs. Max. 3.7 DW	12
Stargrass, *Eleusine indica*	11.3 DW vs. 6.7 DW	12
Water	5.2 (2.8–10.4) DW vs. no data (USEPA drinking water criterion, <0.01 FW)	12
Active gold mining town and environs; 14 sites; 1992–1993		
Soil	12.9 (2.1–48.9) DW	13
Plantain, edible portions	Max. 4.3 DW	13
Water fern, *Ceratopterus cornuta*; whole	9.1 (0.5–78.7) DW	13
Elephant grass, *Pennisetum purpureum*; whole	Max. 27.4 DW	13
Cassava, edible portions	Max. 2.6 DW	13
Mudfish, *Heterobranchus bidorsalis*; whole	Max. 2.7 DW	13

Table 1. (Continued).

Location, sample, and other variables	Concentration (mg total arsenic/ kg DW or FW)[a]	Reference[b]
Tanzania; Serengeti National Park; drainage water from Lake Victoria goldfield tailings	324 FW	14
Europe		
Poland and Czech Republic; 5 species of aquatic bryophytes collected spring-summer		
Ten sites draining an area with high arsenic mineralization	3.4 DW	15
Two sites as above in areas of former gold mining activities	19.4 DW	15
Twenty-two reference sites	0.8 DW	15
Malaysia		
Tributary that received gold mine effluents for at least 10 yr		
Sediments	147 DW	16
Bivalve molluscs; 3 species; soft parts; from sediments containing 6.3 mg As/kg DW (plus, in mg/kg DW, 3.4 Cu, 0.02 Hg, 0.7 Pb, and 27 Zn); no bivalves found in more heavily contaminated sediments	Max. 225 DW (plus 115 mg Cu/kg DW, 127 mg Zn/kg DW, and negligible concentrations of Cd, Pb, and Hg)	16

[a]Ranges in parentheses.
[b]References: 1, Matschullat et al. 2000; 2, Grosser et al. 1994; 3, Tarras-Wahlberg et al. 2000; 4, Da Rosa and Lyon 1997; 5, Azcue et al. 1994; 6, Wong et al. 1999; 7, Bright et al. 1994; 8, Bright et al. 1996; 9, Cain et al. 1992; 10, Custer et al. 2002; 11, Golow et al. 1996; 12, Amonoo-Neizer and Amekor 1993; 13, Amonoo-Neizer et al. 1996; 14, Bowell et al. 1995; 15, Samecka-Cymerman and Kempers 1998; 16, Lau et al. 1998.

one case, gold was extracted by underground mining between 1933 and 1964 near a lake located in northeastern British Columbia, leaving tailings and waste rock 4.5 m thick over 25 ha of land adjacent to the lake. The tailings contained >2000 mg As/kg, the lake sediments up to 1104 mg As/kg, and lake water up to 556 µg/L. The greatest proportion of arsenic in the sediment cores is associated with iron oxides and sulfides. Under aerobic conditions, the high concentrations of iron in the tailings were effective at limiting arsenic migration (Azcue et al. 1994).

Abnormally high concentrations of arsenic in sediment [maximum, 3090 mg As/kg dry weight (DW)] and water samples were documented in 1990–1991 from a watershed receiving gold mine effluent near Yellowknife, Northwest Territories, Canada (Bright et al. 1994, 1996). Inorganic arsenic concentrations were maximal in the water column, sediment particulates, and sediment pore water about 4–6 km downstream of the gold mine input. Arsenite (As^{3+}) was the predominant arsenical in sediment pore water, and arsenate (As^{5+}) was the primary dissolved arsenic species in water column samples. Water samples also contained a variety of methylated arsenicals; methylation of As^{3+} and As^{5+} compounds through biological and other processes reduces their toxicity. Particulate concentrations of arsenic constituted up to 70% of the total arsenic in the water column downstream of the gold mine discharge. The high concentrations of arsenicals in sediment pore water (maximum, 5.16 mg/L) and the overlying water (maximum, 547 µg/L) in dissolved form in areas distant from the input are attributable to remobilization from sediments through redox-related dissolution (Bright et al. 1994, 1996).

Soil contamination by gold mining operations tends to be localized and, because of the phytotoxic effects of arsenic, not easily overlooked (O'Neill 1990). At Yellowknife, Canada, high concentrations of arsenic were measured in soils near a gold smelter: >21,000 mg/kg DW soil at 0.28 km from the smelter, and 600 mg As/kg DW at a site 1 km distant. The tailings deposit also caused contamination of surrounding soils. Vegetation that grew in these contaminated areas usually contained low concentrations of arsenic, except when soil levels were >1000 mg As/kg, which either produced phytotoxic effects in sensitive species or growth in a few tolerant genotypes.

Maximum acceptable concentrations of arsenic in soils used for food production or parks range between 10 and 40 mg As/kg DW in Europe and the United Kingdom (O'Neill 1990). Galbraith et al. (1995) aver that soil arsenic concentrations in excess of 20–50 mg/kg are injurious to plant growth and development and that sensitive species may be affected by concentrations as low as 5 mg/kg; greater levels of these concentrations can lead to toxic responses that include root plasmolysis, necrosis of leaf tips, and seed germination failure. In arsenic-enriched areas, evergreen forests were replaced with bare ground devoid of vegetation, grasslands were dominated by weeds, and there was overall species impoverishment, including wildlife species (Galbraith et al. 1995). Phytoremediation of gold mining sites contaminated by arsenic using arsenic-tolerant plants, such as *Equisetum* spp., is recommended (Wong et al. 1999).

Arsenic contamination in Whitewood Creek, South Dakota, from a gold mine was assessed in aquatic insects and bed sediments over a 40-km reach (Cain et al. 1992). From 1876 to 1977, about 100 million t of finely ground gold mine tailings was discharged via a small tributary into Whitewood Creek; the main contaminant was arsenic derived from arsenopyrites (May et al. 2001). Transport and deposition of the discharged tailings caused extensive downstream arsenic contamination of sediments and biota (Cain et al. 1992). In spring 1987, the maximum arsenic concentration in sediments was 764 mg/kg DW, versus 18 for a reference site. For four species of aquatic insects, the maximum value was 625 mg As/kg DW, versus 16 for a reference site, with most arsenic concentrated in the exoskeleton (Cain et al. 1992). Insectivorous birds (house wren, *Troglodytes aedon*) feeding on these same species of aquatic insects near Whitewood Creek in 1997 had elevated arsenic concentrations in liver (maximum, 5.6 mg As/kg DW) compared to a reference site in Wyoming (<0.5 mg As/kg DW) (Custer et al. 2002).

In Ghana, where gold accounts for the largest proportion of foreign exchange, large quantities (17 t daily) of arsenic are discharged into the atmosphere from a single roasting/smelting facility (Amonoo-Neizer and Amekor 1993). Total arsenic, pentavalent arsenate (As^{5+}), and trivalent arsenite (As^{3+}), were usually highest in soils near the gold ore processing facility (Table 1), with background levels reaching 7–15 km from the site, depending on wind direction and velocity (Golow et al. 1996). Freshwaters in the vicinity of the smelter had grossly elevated concentrations of arsenic (mean, 5.2 mg As/L; range, 2.8–10.4 mg/L; see Table 1), and were considered unfit for aquatic life, irrigation, or human consumption.

In Malaysia, edible clams and mussels from a tributary receiving gold mine wastes contained up to 225 mg As/kg DW soft parts, a level that exceeded mandatory levels for arsenic set by the Malaysian Food Act of 1983 (Lau et al. 1998). Because arsenic enhances the toxicity of free cyanide to aquatic fauna (Leduc 1984), this knowledge needs to be incorporated into future arsenic risk assessments.

V. Arsenic Effects on Sensitive Species

Adverse effects of various arsenicals on sensitive species of organisms have been documented (Table 2) (Eisler 2000). The most sensitive of the aquatic organisms tested showing adverse effects were three species of marine algae, with reduced growth evident in the range of 19–22 µg As^{3+}/L; developing embryos of the narrow-mouthed toad (*Gastrophryne carolinensis*), of which 50% were dead or malformed in 7 d at 40 µg As^{3+}/L; and a freshwater alga (*Scenedesmus obliquis*), in which growth was inhibited 50% in 14 d at 48 µg As^{5+}/L. Adverse biological effects have also been documented at 75–100 µg As/L: growth reduction in freshwater and marine algae at 75 µg As^{5+}/L; 10%–32% mortality in 28 d of a freshwater amphipod (*Gammarus pseudolimnaeus*) at 85–88 µg/L As^{5+} or various methylated arsenicals; inhibition of sexual reproduction of marine algae at 95 µg As^{3+}/L; and death of marine copepods and impaired swimming ability of goldfish at 100 µg As^{5+}/L (Table 2) (Eisler 2000).

Juvenile tanner crabs (*Chionoecetes bairdi*) held for 502 d on weathered gold mine tailings with elevated arsenic concentrations (29.7 mg As/kg DW) or reference sediments (2.5 mg As/kg DW) showed the same concentrations of arsenic in gill (8.9 vs. 9.8 mg As/kg DW) and muscle (8.9 vs. 8.1 mg As/kg DW) tissues (Stone and Johnson 1997). Female tanner crabs may initially avoid areas affected by submarine tailings but later recolonize the altered sea floor and incorporate lead, but not arsenic, into their tissues (Stone and Johnson 1998). In a 90-d study of ovigerous tanner crabs in forced contact with fresh gold mine tailings, survival and reproduction were normal, although egg survival was lower than in those held on control sediments, which was attributed to the action

Table 2. Lethal and sublethal effects of various arsenicals on selected species of plants, animals, and humans.

Taxonomic group, species, arsenic compound, and dose	Effect	Ref[a]
Freshwater Plants		
Algae; 4 species; As^{3+} (inorganic trivalent arsenite); 1.7–2.3 mg/L	95%–100% fatal in 2–4 wk	1, 2
Algae; As^{5+} (inorganic pentavalent arsenate); 2 species; 0.048–0.26 mg/L	50% growth inhibition in 14 d	2
Freshwater Invertebrates		
Cladocerans		
Bosmina longirostris; As^{5+}; 0.85 mg/L	50% immobilization in 96 hr	3
Daphnia magna		
As^{5+}; 0.52 mg/L	16% reproductive impairment in 3 wk	2
As^{3+}; 0.63–1.32 mg/L	MATC[b]	2
As^{5+}; 7.4 mg/L	50% dead in 96 hr	1
Daphnia pulex		
As^{3+}; 1.3 mg/L	50% dead in 96 hr	1, 2
As^{3+}; 3.0 mg/L	50% immobilized in 48 hr	4
As^{5+}; 49.6 mg/L	50% immobilized in 48 hr	3
Simocephalus serrulatus; As^{3+}; 0.81 mg/L	50% dead in 96 hr	2
Amphipod, *Gammarus pseudolimnaeus*		
DSMA, disodium methylarsenate $[CH_3AsO(ONa)_2]$; 0.086 mg/L	10% dead in 28 d	5
As^{3+}; 0.088 mg/L	20% dead in 28 d	5
SDMA, sodium dimethylarsenate $[(CH^3)_2As(ONa)]$; 0.85 mg/L	No deaths in 28 d	5
As^{3+}; 0.96 mg/L	All dead in 28 d	5
As^{5+}; 0.97 mg/L	20% dead in 28 d	5
DSMA; 0.97 mg/L	40% dead in 28 d	5
Snail, *Helisoma campanulata*		
SDMA; 0.085 mg/L	No deaths in 28 d	5
As^{3+}; 0.96 mg/L	10% dead in 28 d	5
As^{5+}; 0.97 mg/L	No deaths in 28 d; maximum bioconcentration factor = 99	5
DSMA; 0.97 mg/L	No deaths in 28 d	5
Red crayfish, *Procambarus clarki*		
MSMA, monosodium methanearsonate $[CH_4AsNaO_3]$; 100 mg/L, equivalent to 46.3 mg As/L	No effect on growth or survival during exposure for 24 wk, but hatching success reduced to 17% vs. 78% for controls	6
MSMA; 1,000 mg/L	50% dead in 96 hr	6

Table 2. (Continued).

Taxonomic group, species, arsenic compound, and dose	Effect	Ref[a]
Stoneflies		
Pteronarcys californica; As^{3+}; 38.0 mg/L	50% dead in 96 hr	4
Pteronarcys dorsata		
DSMA, SDMA, or As^{3+}; 0.85–0.97 mg/L	No deaths in 28 d	5
As^{5+}; 0.97 mg/L	20% dead in 28 d	5
Freshwater Fishes		
Goldfish, Carassius auratus; As^{5+}; 0.1 mg/L	15% behavioral impairment in 24 hr; 30% impairment in 48 hr	7
Flagfish, Jordanella floridae		
As^{3+}; 2.1–4.1 mg/L	$MATC^b$	2
As^{3+}; 14.4 mg/L	50% dead in 96 hr	8
Fathead minnow, Pimephales promelas		
As^{5+}; 0.53–1.50 mg/L	$MATC^b$	2
As^{3+}; 2.1–4.8 mg/L	$MATC^b$	8
As^{3+}; 14.1 mg/L	50% dead in 96 hr	8
As^{5+}; 25.6 mg/L	50% dead in 96 hr	2
Rainbow trout, Oncorhynchus mykiss		
As^{3+}; 0.54 mg/L	Embryos: 50% dead in 28 d	1
DSMA or SDMA; 0.85–0.97 mg/L	No deaths in 28 d	5
As^{3+}; 0.96 mg/L	50% dead in 28 d	4, 9
As^{3+}; 23.0–26.6 mg/L	Adults: 50% dead in 28 d	5
Sodium cacodylate (SC); 1,000 mg/L	No deaths in 28 d	11
As^{5+}; 10–90 mg/kg diet for 16 wk	No effect level at ~ 10 mg/kg diet; some adaptation to 90 mg/kg diet as initial negative growth gave way to slow positive growth over time	10
DSA, disodium arsenate heptahydrate; 13–33 mg As as DSA/kg ration for 12-24 wk (0.28–0.52 mg As/kg BW/d	$MATC^b$	5
As^{3+} or As^{5+}; 120–1,600 mg/kg diet for 8 wk	Growth depression, food avoidance, and impaired feed efficiency at all levels	10
DMA, dimethyl arsinic acid, or ABA, p-amino-benzene-arsonic acid; 120–1,600 mg/kg diet for 8 wk	No toxic response at any level tested	10
Amphibians		
Marbled salamander, Ambystoma opacum; As^{3+}; 4.5 mg/L	Developing embryos: 50% dead or malformed in 8 d	2
Narrow-mouthed toad, Gastrophryne carolenisis; As^{3+}; 0.04 mg/L	Developing embryos: 50% dead or malformed in 7 d	2
Marine Plants		
Algae, 3 species; As^{3+}; 0.019–0.022 mg/L	Reduced growth	2

Table 2. (Continued).

Taxonomic group, species, arsenic compound, and dose	Effect	Ref[a]
Red alga, *Champia parvula*		
As^{3+}; 0.065 mg/L	Normal sexual reproduction	12
As^{3+}; 0.095 mg/L	No sexual reproduction	12
As^{3+}; 0.300 mg/L	Death	12
As^{5+}; 10.0 mg/L	Normal growth but no sexual reproduction	12
Phytoplankton; As^{5+}; 0.075 mg/L	Reduced biomass of populations in 4 d	2
Alga, *Skeletonema costatum*; As^{5+}; 0.13 mg/L	Growth inhibition	2
Marine Invertebrates		
Copepod, *Acartia clausi*; As^{3+}; 0.51 mg/L	50% dead in 96 hr	2
Copepod, *Eurytermora affinis*		
As^{5+}; 0.1 mg/L	Reduced juvenile survival	13
As^{5+}; 1.0 mg/L	Reduced adult survival	13
Dungeness crab, *Cancer magister*; As^{3+}; 0.23 mg/L	Zoeae: 50% dead in 96 hr	2
Pacific oyster, *Crassostrea gigas*; As^{3+}; 0.33 mg/L	Embryos: 50% dead in 96 hr	2
Mysid, *Mysidopsis bahia*		
As^{3+}; 0.63–1.27 mg/L	MATC[b]	2
As^{5+}; 2.3 mg/L	50% dead in 96 hr	2
Marine Fishes		
Three species; As^{3+}; 12.7–16.0 mg/L	50% dead in 96 hr	2
Pink salmon, *Oncorhynchus gorbuscha*		
As^{3+}; 2.5 mg/L	No deaths in 10 d	9
As^{3+}; 3.8 mg/L	54% dead in 10 d	2
As^{3+}; 7.2 mg/L	All dead in 7 d	2
Terrestrial Plants		
Crops		
Total water-soluble arsenic; 3–28 mg/kg soil	Depressed crop yield	7
Total arsenic; 25–85 mg/kg soil	Depressed crop yield	7
Common bermudagrass, *Cynodon dactylon*; As^{3+}; arsenic-amended soils containing up to 90 mg As/kg soil	Arsenic residues were up to 17 mg/kg DW in stems, 20 in leaves, and 304 in roots	14
Soybean, *Glycine max*; total arsenic; >1 mg/kg DW plant	Toxic signs	7
Rice, *Oryza sativa*; DSMA; 50 mg/kg soil	75% decrease in yield	7
Scots pine, *Pinus sylvestris*		
As^{5+}; >62 mg/kg shoots DW	Toxic	15

Table 2. (Continued).

Taxonomic group, species, arsenic compound, and dose	Effect	Ref[a]
As^{5+}; >250 mg/kg soil DW	Seedlings die	15
As^{5+}; >3,300 mg/kg shoots DW	Fatal	15
Pea, *Pisum sativum*; As^{3+}; 15 mg/L	Inhibition of light activation and photosynthetic CO_2 fixation in chloroplasts	19
Grasslands; CA, cacodylic acid [$(CH_3)_2$ AsO(OH)]; 17kg/ha	75%–90% of all species killed; recovery modest	11
Sandhill plant communities		
CA; 2.25 kg/ha	No lasting effect	11
CA; 6.8 kg/ha	Some species defoliated	11
CA; 34.0 kg/ha	75% defoliation of oaks and death of all pine trees	11
Terrestrial Invertebrates		
Beetles; CA; dietary levels 100–1,000 mg/kg	Fatal to certain pestiferous species	16
Western spruce budworm, *Christoneura occidentalis*; sixth-instar larvae		
As^{3+}; 99.5 mg/kg ration fresh weight (FW)	Fatal to 10%	17
As^{3+}; 2,250 mg/kg ration FW	Fatal to 50%	17
As^{3+}; 65,300 mg/kg ration FW	Fatal to 90%	17
As^{3+}; 100–65,300 mg/kg ration FW	Newly molted pupae and adults of As-exposed larvae had reduced weight; regardless of dietary levels, concentrations of As ranged up to 2,640 mg/kg DW in dead pupae and 1,708 mg/kg DW in adults	17
Earthworm, *Lumbricus terrestris*		
As^{5+}; 40 mg/kg dry weight (DW) soil; exposure for 23 d	No accumulations in first 12 d, bioconcentration factor (BCF) = 3 by d 23	18
As^{5+}; 100 mg/kg DW soil	Fatal to 50% in 8 d	18
As^{5+}; 400 mg/kg DW soil	Fatal to 50% in 2 d	18
Birds		
Mallard, *Anas platyrhynchos*		
Adult breeding pairs; As^{5+}; fed diets with 0, 25, 100, or 400 mg/kg ration up to 173 d; ducklings produced were fed the same diet as their parents for 14 d	Dose-dependent increase in liver arsenic from 0.23 mg As/kg DW in controls to 6.6 in the 400 mg/kg group and in eggs from 0.23 in controls to 3.6 mg/kg DW in the 400 mg/kg group. Dose-dependent adverse effects on growth, onset of egg laying, and eggshell thinning. In ducklings, arsenic accumulated in the liver from 0.2 mg As/kg DW in controls to 33.0 in the 400 mg/kg group and caused a dose-dependent decrease in growth rate of whole body and liver.	20

Table 2. (Continued).

Taxonomic group, species, arsenic compound, and dose	Effect	Ref[a]
Ducklings; As^{5+}; fed 30, 100, or 300 mg/kg diet for 10 wk	All treatments produced elevated hepatic glutathione and ATP concentrations and decreased overall weight gain and rate of growth in females. Arsenic concentrations were elevated in brain and liver of ducklings fed 100 or 300 mg/kg diet; all ducklings had altered behavior, e.g., increased resting time; males had reduced growth.	21
Day-old ducklings; As^{5+}; fed diets containing 200 mg/kg ration for 4 wk	When protein was adequate (22%), some growth reduction resulted, with only 7% protein in diet, growth and survival were reduced and frequency of liver histopathology increased	22
Adult males; As^{5+}; fed rations containing 300 mg/kg	Equilibrium reached in 10–30 d; 50% loss from liver in 1–3 days on transfer to an uncontaminated diet	23
As^{3+}; 323 mg As^{3+}/kg body weight (BW)	Acute oral LD_{50}	7, 9, 24
As^{3+}; 500 mg/kg diet	Fatal to 50% in 32 d	2
As^{3+}; 1,000 mg/kg diet	Fatal to 50% in 6 d	2
CA; 1,740–5,000 mg/kg diet	Fatal to 50% in 5 d	11
California quail, *Callipepla californica*; As^{3+}; 47.6 mg/kg BW	Acute oral LD_{50}	24
Common bobwhite, *Colinus virginianus*		
SC; 1,740 mg/kg diet for 5 d	No effects on behavior; no signs of intoxication; negative necropsy	11
MSMA; 3,300 mg/kg BW	Acute oral LD_{50}	11
Chicken, *Gallus gallus*		
As^{3+}; 0.01–1.0 mg/embryo	Up to 34% dead; malformation threshold at 0.03–0.3 mg/embryo	7
As^{5+}; 0.01–1.0 mg/embryo	Up to 8% dead	7
As^{5+}; 0.3–3.0 mg/embryo	Malformation threshold	7
DSMA; 1–2 mg/egg	Teratogenic when injected	7, 11
SC; 1–2 mg/egg	Developmental abnormalities when injected	11
DC, dodecylamine *p*-chlorophenylarsonate; 23.3 mg/kg diet for 9 wk	Liver residues were 2.9 mg/kg FW at end; no ill effects noted	25
CA; 100 mg/kg BW	No adverse effects at daily oral dosing for 10 days	11

Table 2. (Continued).

Taxonomic group, species, arsenic compound, and dose	Effect	Ref[a]
Ring-necked pheasant, *Phasianus colchicus*; As^{3+}; 363 mg/kg BW	Acute oral LD_{50}	24
Mammals		
Cattle, *Bos* spp.		
As^{5+}; fed 33 mg daily per animal for 33 mon	Elevated levels of arsenic in muscle (0.02 mg/kg FW vs. 0.005 in controls) and liver (0.03 vs. 0.012) but normal levels in milk and kidney	26
As^{3+}; fed 33 mg daily per animal for 15–28 mon	Elevated arsenic levels, in mg/kg FW, of 0.002 for milk (vs. 0.001 for controls), 0.03 for muscle (vs. 0.005), 0.1 for liver (vs. 0.012), and 0.16 for kidney (vs. 0.053)	26
As^{3+}; single oral dose 15–45 g/animal, as arsenic trioxide	Fatal	7
As^{3+}; single oral dose 1–4 g/animal, as sodium arsenite	Fatal	7
MSMA; 10 mg/kg BW daily for 10 d	Fatal	7
As^{3+}; 33–55 mg/kg BW, or 13.2–22 g for a 400-kg animal; topical application	Arsenic-poisoned cows contained up to 15 mg As/kg FW liver, 23 in kidney, and 45 in urine (vs. <1 for all normal tissues)	27
CA or MAA (methanearsonic acid); calves fed diets containing 4,000–4,700mg/kg ration	Anoretic in 3–6 d	11
CA: adults given oral dose of 10 mg/kg BW daily for 3 wk, followed by 20 mg/kg BW daily for 5–6 wk	Lethal	11
CA; adults given oral dose of 25 mg/kg BW daily for 10 d	Adverse effects	11
Dog, *Canis familiaris*		
CA or MMA; 30 mg/kg diet for 90 d	No adverse effects	11
CA; 1,000 mg/kg BW	Oral LD_{50}	11
As^{3+}; 50–150 mg	Fatal	7
Guinea pig, *Cavia* sp.; As^{3+} as arsenic trioxide; fed diet containing 50 mg/kg for 21 d	Elevated arsenic residues (mg/kg FW), 4 in blood and 15 in heart vs. <1 for controls	25
Hamster, *Cricetus* sp.		
As^{5+}; maternal dose 5 mg/kg BW	Some fetal deaths, but no malformations	7

Table 2. (Continued).

Taxonomic group, species, arsenic compound, and dose	Effect	Ref[a]
As^{5+}; maternal dose 20 mg/kg BW	54% fetal deaths and malformations	7
As^{5+}, as sodium arsenate; dosed intravenously on day 8 of gestation		
2 mg/kg BW	No measurable effect	28
8 mg/kg BW	Increased incidence of malformation and resorption	28
16 mg/kg BW	All embryos died	28
SC; single intraperitoneal injection 900–1,000 mg/kg BW during midgestation	Some maternal deaths and increased incidences of fetal malformations	11
Horse, *Equus caballus*; As^{3+}; 2–6 mg/kg BW daily (1–3 g sodium arsenite)	Fatal in 14 wk	7
Cat, *Felis domesticus*; As^{3+} or As^{5+}; 1.5 mg/kg BW daily	Chronic oral toxicity	28
Mammals, representative species		
Calcium arsenate; 35–1,000 mg/kg BW	Single oral LD_{50} range	7
Lead arsenate; 10–50 mg/kg BW	Single oral LD_{50} range	7
As^{3+}, as arsenic trioxide; 3–250 mg/kg BW	Lethal	9
As^{3+}, as sodium arsenite; 1–25 mg/kg BW	Lethal	9
Mouse, *Mus* spp.		
As^{5+}; maternal dose 10 mg/kg BW	Some fetal deaths and malformations	7
As^{5+}; 20–50 mg/kg BW; pregnant mice, d 18 of gestation	No deaths or abortions at lower dose when administered intraperitoneally, or higher dose when given orally; residue half-life was 10 hr regardless of route	32
As^{3+}, as arsenic trioxide		
10.4 mg/kg BW	No deaths in 96 hr after single oral dose	9
39.4 mg/kg BW	Oral LD_{50} in 96 hr	9
0.26 mg/m^3 air for 4 hr daily on days 9–12 of gestation	3.1% decrease in fetal weight	29
2.9 mg/m^3 air for 4 hr daily on days 9–12 of gestation	9.9% decrease in fetal weight	29
28.5 mg/m^3 air for 4 hr daily on days 9–12 of gestation	Fetotoxic effects (reduced survival, impaired growth, retarded limb ossification, bone abnormalities) and chromosomal damage to liver cells by day 18	29

Table 2. (Continued).

Taxonomic group, species, arsenic compound, and dose	Effect	Ref[a]
As^{5+}, as sodium arsenate; 0.5 mg/L drinking water or up to 26 mon, equivalent to 0.07–0.08 mg/kg BW daily	No tumors in controls vs. 41.1% of mice in treated groups with one or more tumors, mostly of the lung, liver, and GI tract	33
As^{3+}, as sodium arsenite; 5 mg/kg diet for three generations	Reduced litter size, but outwardly normal	28
As^{3+}, as sodium arsenite; 9.6–11.3 mg/kg BW via subcutaneous injection	Lower dose is LD_{50}; higher dose is LD_{90} 7 d postexposure	34
As^{3+}, as sodium arsenite; 10–12 mg/kg BW via intraperitoneal route	Lower dose causes damage to bone marrow and sperm; higher dose is LD_{50}	35
Single oral dose		
Arsenous oxide; 34 mg As/kg BW	LD_{50}	31
Tetramethylarsonium iodide; 890 mg As/kg BW	LD_{50}	31
Arsenocholine; 6,500 mg As/kg BW	LD_{50}	31
Arsenobetaine; >100,000 mg As/kg BW	LD_{50}	31
DMA: 200–600 mg/kg BW daily for 10 d	Fetal and maternal toxicity	30
CA; oral dosages 400–600 mg/kg BW on days 7–16 of gestation	Fetal malformations (cleft palate), delayed skeletal ossification, and fetal weight reduction	11
SC; 1,200 mg/kg BW during midgestation via intraperitoneal injection	Increased rates of fetal skeletal malformations	11
Rabbit, *Oryctolagus* sp.; MMA; 50 mg/kg ration for 7–12 wk	Hepatotoxicity	30
Domestic sheep, *Ovis aries*		
As^{3+}, as sodium arsenite; single oral dose 5–12 mg/kg BW (0.2–0.5 g)	Acutely toxic	7
As^{5+}, as soluble arsenic; lambs fed diets containing 2 mg As/kg supplemental arsenic for 3 mon	Maximum arsenic concentrations (mg/kg FW), 2 in brain (vs. 1 in controls), 14 in muscle (2), 24 in liver (4), and 57 in kidney (10)	36
Total arsenic; diets contained lakeweed (*Lagarosiphon major*; 288 mg As/kg DW) at 58 mg total As/kg diet for 3 wk	No ill effects; tissue residues increased during feeding, but rapidly declined when lakeweed was removed from diet	25

Table 2. (Continued).

Taxonomic group, species, arsenic compound, and dose	Effect	Ref[a]
Rat, *Rattus* spp.		
Arsanilic acid; 17.5 mg/kg diet for seven generations	No teratogenesis observed; positive effect on litter size and survival	28
As^{5+}; fed diets containing 50 mg/kg for 10 wk	No effect on serum uric acid levels	37
As^{3+}, as arsenic trioxide; single oral dose 15.1 mg/kg BW	LD_{50} (96 hr)	9
As^{3+}, as arsenic trioxide; fed diets with 50 mg/kg for 21 d	Tissue arsenic levels elevated in blood (125 mg/L vs. 15 in controls), heart (43.0 mg/kg FW vs. 3.3), spleen (60.0 vs. 0.7), and kidney (25.0 vs. 1.5)	25
As^{3+}; oral administration of 1.2 mg/kg BW daily for 6 wk	Serum uric acid levels reduced 67%	37
As^{3+}; 10 mg/L in drinking water for 7 mon	Urinary metabolites were mainly methylated arsenic metabolites with about 6% in inorganic form	38
Arsenobetaine; 100 mg As/L drinking water for 7 mon	Eliminated in urine unchanged without transformation	38
Cacodylic acid (CA); pregnant rats dosed by gavage at 50–60 mg/kg BW daily during gestation days 6–13	Maternal deaths and fetal deaths and abnormalities noted	11
Dimethylarsinic acid (DMA); 100 mg/L in drinking water for 7 mon	Main metabolites in urine were DMA and trimethylarsin oxide (TMAO) with minute amounts of tetramethylarsonium (TMA)	38
DMA; 40–60 mg/kg BW daily for 10 d	Fetal and maternal toxicity	30
Monomethylarsonic acid (MMA); 200 mg/L in drinking water for 7 mon	Main products in urine were unchanged MMA, DMA, and small amounts of TMA and TMAO	38
Rodents, various species		
Cacodylic acid (CA); 470–830 mg/kg BW	LD_{50} range by various routes of administration	11
Sodium cacodylate (SC); 600–2,600 mg/kg BW	LD_{50} range, various routes of administration	
Cotton rat, *Sigmodon hispidus*; As^{3+} as sodium arsenite; adult males given 0, 5, or 10 mg/L in drinking water for 6 wk	Dose-dependent decrease in daily food intake; minimal effects on immune function, tissue weights, and blood chemistry	39

Table 2. (Continued).

Taxonomic group, species, arsenic compound, and dose	Effect	Ref[a]
Pig, *Sus* sp.		
As^{3+}, as sodium arsenite; 500 mg/L in drinking water	Lethal when arsenic residues 100–200 mg/kg BW	9
3-Nitro-4-hydroxyphenyl-arsonic acid; 100–250 mg/kg diet	Arsenosis documented after 2 mon on diets containing 100 mg/kg or after 3–10 d on diets containing 250 mg/kg	9
Human Health		
As^{5+}; 3.5 mg daily for 1 mon	12,000 Japanese infants accidentally poisoned (128 deaths) from consumption of dry milk contaminated with arsenic; postexposure effects (15 yr later) included severe hearing loss, brain wave abnormalities, and other CNS disturbances	28
As^{3+} as arsenic trioxide		
1–2.6 mg/kg BW (70–189 mg)	Some deaths	7
7 mg/kg BW	LD_{50}	7
CA; 1350 mg/kg BW	LD_{50}	7
Total arsenic; 1–3 mg/kg BW daily for 3 mon in children or 80 mg kg/BW daily for 3 mon in adults	Symptoms of chronic arsenic poisoning	7
Total arsenic in drinking and cooking water; prolonged use		
0.29 mg/L	Skin cancer	7
0.6 mg/L	Chronic arsenic intoxication	7
Total inorganic arsenic; 3 mg daily for 2 wk	May cause severe poisoning in infants and symptoms of toxicity in adults	28

[a]References: 1, USEPA 1980; 2, USEPA 1985; 3, Passino and Novak 1984; 4, Johnson and Finley 1980; 5, Spehar et al. 1980; 6, Naqvi and Flagge 1990; 7, NRCC 1978; 8, Lima et al. 1984; 9, NAS 1977; 10, Cockell and Hilton 1985; 11, Hood 1985; 12, Thursby and Steele 1984; 13, Sanders 1986; 14, Wang et al. 1984; 15, Sheppard et al. 1985; 16, Jenkins 1980; 17, Robertson and McLean 1985; 18, Meharg et al. 1998; 19, Marques and Anderson 1986; 20, Stanley et al. 1994; 21, Camardese et al. 1990; 22, Hoffman et al. 1992; 23, Pendleton et al. 1995; 24, Hudson et al. 1984; 25, Woolson 1975; 26, Vreman et al. 1986; 27, Robertson et al. 1984; 28, Pershagen and Vahter 1979; 29, Nagymajtenyi et al. 1985; 30, Hughes and Kenyon 1998; 31, Hamasaki et al. 1995; 32, Hood et al. 1987; 33, Ng et al. 1998; 34, Stine et al. 1984; 35, Deknudt et al. 1986; 36, van der Veen and Vreman 1986; 37, Jauge and Del-Razo 1985; 38, Yoshida et al. 1998; 39, Savabieasfahani et al. 1998.

[b]MATC, Maximum acceptable toxicant concentration. Lower value in each pair indicates highest concentration tested producing no measurable effect on growth, survival, reproduction, or metabolism during chronic exposure; higher value indicates lowest concentration tested producing a measurable effect.

of lead; arsenic concentrations in muscle and ova were similar for those held on control and tailings sediments (Stone and Johnson 1998). Reduced food availability to ovigerous females due to smothering of the sea floor could result in reduced fecundity, poor larval survival, and increased susceptibility to disease (Johnson et al. 1998b).

Juvenile yellowfin sole (*Pleuronectes asper*) avoid fresh tailings (15 mg As/kg DW) in favor of natural marine sediments (7 mg As/kg DW), but when tailings are covered with 2 cm control sediments, there is no significant avoidance of the covered fresh tailings (Johnson et al. 1998a). Growth was inhibited for sole held on fresh tailings for 30 d but not during days 30 to 60; survival was similar (90%–93% survival) for fish held on all sediments (Johnson et al. 1998a).

Among terrestrial plants and invertebrates, yields of most crops decreased at soil arsenic levels of 3–28 mg water-soluble arsenic/L and 25–85 mg/kg of total arsenic; yields of peas (*Pisum sativum*) were decreased at 1 mg/L water-soluble arsenic or 25 mg/kg total soil arsenic; soybeans (*Glycine max*) grew poorly when plant residues exceeded 1 mg As/kg DW; and earthworms (*Lumbricus terrestris*) held in soils containing 40–100 mg As^{5+}/kg DW soil for 23 d showed reduced survival, especially among worms held in soils <70 mm in depth when compared to worms held at 500–700 mm (Table 2; Eisler 2000).

Signs of inorganic trivalent arsenite poisoning in birds (muscular incoordination, debility, slowness, jerkiness, falling, hyperactivity, fluffed feathers, drooped eyelid, huddled position, unkempt appearance, loss of righting reflex, immobility, seizures) were similar to those induced by many other toxicants and did not seem to be specific for arsenosis. Signs occurred within 1 hr and deaths within 1–6 d postadministration; remission took up to 1 mon (Hudson et al. 1984). Internal examination suggested that lethal effects of acute inorganic arsenic poisoning were caused by the destruction of the blood vessels lining the gut, which resulted in decreased blood pressure and subsequent shock (Nystrom 1984). Mallard ducklings fed a diet of 30-mg As/kg ration had reduced growth, and those fed a 300-mg As/kg diet had altered brain biochemistry and nesting behavior (see Table 2). Decreased energy levels and altered behavior can further decrease duckling survival in a natural environment (Camardese et al. 1990).

In mammals, arsenic uptake may occur by ingestion (the most likely route), inhalation, and absorption through the skin and mucous membranes. Soluble arsenicals are absorbed more rapidly and completely than are the sparingly soluble arsenicals, regardless of route of administration [National Research Council of Canada (NRCC) 1978]. In humans, inorganic arsenic at high concentrations is associated with adverse reproductive outcomes, including increased rates of spontaneous abortion, low birth weight, congenital malformations, and death (Hopenhayn-Rich et al. 1998). However, at environmentally relevant levels and routes of exposure, humans are not at risk for birth defects due to arsenic (Holson et al. 1998). *In vitro* tests with human erythrocytes demonstrate that inorganic As^{5+} as sodium arsenate was as much as 1000 times more effective than inorganic As^{3+} as sodium arsenite after exposure to 750 mg As/L in causing death, morphological changes, and ATP depletion (Winski and Carter 1998).

Acute episodes of poisoning in warm-blooded organisms by inorganic and organic arsenicals are usually characterized by high mortality and morbidity over a period of 2–3 d [National Academy of Sciences (NAS) 1977; Selby et al. 1977]. General signs of arsenic toxicosis include intense abdominal pain, staggering gait, extreme weakness, trembling, salivation, vomiting, diarrhea, fast and feeble pulse, prostration, collapse, and death. Gross necropsy shows a reddening of gastric mucosa and intestinal mucosa, a soft yellow liver, and red edematous lungs, Histopathological findings show edema of gastrointestinal mucosa and submucosa, necrosis and sloughing of mucosal epithelium, renal tubular degeneration, hepatic fatty changes and necrosis, and capillary degeneration in the gastrointestinal tract, vascular beds, skin, and other organs.

In subacute episodes, in which animals live for several days, signs of arsenosis include depression, anorexia, increased urination, dehydration, thirst, partial paralysis of rear limbs, trembling, stupor, coldness of extremities, and subnormal body temperatures [NAS 1977; Selby et al. 1977; U.S. Public Health Service (USPHS) 2000]. In cases involving cutaneous exposure to arsenicals, a dry, cracked, leathery, and peeling skin may be a prominent feature (Selby et al. 1977). Nasal discharges and eye irritation were documented in rodents exposed to organoarsenicals in inhalation toxicity tests (Hood 1985). Subacute effects in humans and laboratory animals include peripheral nervous disturbances, melanosis, anemia, leukopenia, cardiac abnormalities, and liver changes. Most adverse signs rapidly disappear after exposure ceases (Pershagen and Vahter 1979).

Research results on arsenic poisoning in mammals (see Table 2; Eisler 2000) show general agreement on eight points:

1. Arsenic metabolism and effects are significantly influenced by the organism tested, the route of administration, the physical and chemical form of the arsenical, and the dose.
2. Inorganic arsenic compounds are more toxic than organic arsenic compounds, and trivalent species are more toxic than pentavalent species.
3. Inorganic arsenicals can cross the placenta in most species of mammals.
4. Early developmental stages are the most sensitive, and humans appear to be one of the more susceptible species.
5. Animal tissues usually contain low levels (<0.3 mg As/kg FW) of arsenic. After the administration of arsenicals, these levels are elevated, especially in liver, kidney, spleen, and lung; several weeks later, arsenic is translocated to ectodermal tissues (hair, nails) because of the high concentration of sulfur-containing proteins in these tissues.
6. Inorganic arsenicals are oxidized *in vivo*, biomethylated, and usually excreted rapidly in the urine, but organoarsenicals are usually not subject to similar transformations.
7. Acute or subacute arsenic exposure can lead to elevated tissue residues, appetite loss, reduced growth, loss of hearing, dermatitis, blindness, degenerative

changes in liver and kidney, cancer, chromosomal damage, birth defects, and death.
8. Death or malformations have been documented at single oral doses of 2.5–33 mg As/kg body weight (BW), at chronic doses of 1–10 mg As/kg BW, and at dietary levels >5 and <50 mg As/kg diet.

Unlike wildlife, reports of arsenosis in domestic animals are common in cattle and house cats, less common in sheep and horses, and rare in pigs and poultry (NAS 1977). In practice, the most dangerous arsenic preparations are dips, herbicides, and defoliants in which the arsenical is in a highly soluble trivalent form, usually as trioxide or arsenite (Selby et al. 1977). Accidental poisoning of cattle with arsenicals, for example, is well documented. In one instance, more than 100 cattle died after accidental overdosing with arsenic trioxide applied topically to control lice. On necropsy, there were subcutaneous edematous swellings and petechial hemorrhages in the area of application, as well as histopathology of the intestine, mucosa, kidney, and epidermis (Robertson et al. 1984).

When extrapolating animal data from one species to another, the species tested must be considered. For example, the metabolism of arsenic in the rat (*Rattus* sp.) is unique and very different from that in humans and other mammals. Rats store arsenic in blood hemoglobin, excreting it slowly, unlike most mammals, which rapidly excrete ingested inorganic arsenic in the urine as methylated derivatives (NAS 1977). Blood arsenic, whether given as As^{3+} or As^{5+}, rapidly clears from humans, mice, rabbits, dogs, and primates; the half-life is 6 hr for the fast phase and about 60 hr for the slow phase [U.S. Environmental Protection Agency (USEPA) 1980]. In the rat, however, blood arsenic is mostly retained in erythrocytes and clears slowly; the half-life is 60–90 d (USEPA 1980). In rats, the excretion of arsenic into bile is 40 times faster than in rabbits and up to 800 times faster than in dogs (Pershagen and Vahter 1979). Most researchers agree that the rat is unsatisfactory for use in arsenic research (NAS 1977; NRCC 1978; Pershagen and Vahter 1979; USEPA 1980; Webb et al. 1986).

VI. Proposed Arsenic Criteria

Numerous arsenic criteria have been proposed for the protection of human health and natural resources; some are shown in Table 3. Most proposed arsenic criteria have been exceeded, sometimes by orders of magnitude, in samples collected near gold mining extraction and refining facilities (see Tables 1, 2).

Arsenic criteria are undergoing constant revision. For example, the criterion of 190 µg As^{3+}/L for freshwater life protection (USEPA 1985) was reduced over a 5-yr period from 440 µg As^{3+}/L (USEPA 1980) but still does not afford adequate protection; many species of freshwater biota are adversely affected at <190 µg/L As^{3+}, As^{5+}, or various organoarsenicals (Table 2). These adverse effects include death and malformations of toad embryos at 40 µg/L, growth inhibition of algae at 48–74 µg/L, mortality of amphipods and gastropods at 85–88

Table 3. Proposed arsenic criteria for the protection of human health and selected natural resources.

Resource and other variables	Criterion or effective arsenic concentration	Ref[a]
Human Health		
Total diet	<0.5 mg As/kg dry weight (DW) diet; 0.0003–0.0008 mg/kg body weight (BW) daily	1, 13
Total intake	No observable effect at <0.021 mg arsenic daily based on 0.0003 mg/kg BW daily for 70-kg adult	1
Muscle of poultry and swine, eggs, swine edible byproducts	<2 mg As/kg fresh weight (FW)	2
Shellfish diet		
Crustaceans, edible tissues	<76 mg total As/kg FW tissue	3
Tolerable daily intake	<0.13 mg	4
Maximum allowable	<30 mg total As/kg FW diet	4
90th percentile consumers of shellfish		
Bivalve molluscs	0.057 mg daily	4
Lobsters, shrimp	0.18 mg daily	4
Drinking water		
Total arsenic, recommended	<10 µg/L	5, 6, 7, 8
Symptoms of arsenic toxicity observed	9% incidence at 50 µg/L, 16% at 50–100 µg/L, 44% at >100 µg/L	9
Cancer frequency	0.01% at 82 µg As/L; 0.17% at 600 µg As/L	9
Tissue Residues		
No observed effect levels	<0.05 mg As/L urine; <0.5 mg/kg liver or kidney; <0.7 mg/L blood; <2 mg/kg hair; <5 mg/kg fingernail	9
Arsenic-poisoned; liver or kidney	2–100 mg As/kg FW	10
Arsenic-poisoned; whole body; children vs. adults	1 mg As/kg BW (equivalent to intake of 10 mg/mon for 3 mon) vs. 80 mg As/kg BW (intake of 2 g/yr for 3 yr)	9
Air		
Inorganic arsenic, occupational vs. residential	<2 µg/m^3 vs. <10 µg/m^3	1
Organic arsenic	<500 µg/m^3	1
Increased mortality	>3 µg/m^3 for 1 yr	9
Respiratory cancer, increased risk	Lifetime occupational exposure >54.6 µg As/m^3; 50 µg As/m^3 for more than 25 yr	9, 11

Table 3. (Continued).

Resource and other variables	Criterion or effective arsenic concentration	Ref[a]
Skin diseases	60–13,000 µg As/m^3	9
Dermatitis	300–81,500 µg As/m^3	9
Soils used for food production or parks in Europe and UK	10–40 mg As/kg DW	12
Terrestrial Vegetation		
No observable effects	<1.0 mg total water-soluble soil As/L, <25.0 mg total As/kg soil, <3.9 µg As/m^3 air	9
Adverse effects, crops and vegetation	3–28 mg water-soluble As/L, equivalent to 25–85 mg total As/kg soil; air concentrations >3.9 µg As/m^3	7
Soils, recommended	<20 mg/kg (Germany) to <500 mg/kg elsewhere	8
Phytotoxic or growth inhibition of tolerant genotypes	>1000 mg/kg DW soil	12
Aquatic Biota		
Freshwater biota: medium	96-hr average water concentration should not exceed 190 µg total recoverable inorganic As^{3+}/L more than once every 3 yr	14
Freshwater biota: tissue residues	Diminished growth and survival in immature bluegills (*Lepomis macrochirus*) when total arsenic residues in muscle >1.3 mg/kg FW or >5 mg/kg in adults	9
Saltwater biota: medium	96-hr average water concentration should not exceed 36 µg As^{3+}/L more than once every 3 yr	14
Saltwater biota: tissues	Depending on chemical form of arsenic, certain marine fishes can tolerate muscle loading of 40 mg total As/kg FW	
Birds		
Single oral dose fatal to 50%; sensitive species	17–48 mg As/kg BW	7
Tissue residues, liver and kidney	Residues of 2–10 mg total As/kg FW are considered elevated; residues >10 mg/kg are indicative of arsenic poisoning	15, 16

Table 3. (Continued).

Resource and other variables	Criterion or effective arsenic concentration	Ref[a]
Diet	Reduced growth in mallard (*Anas platyrhynchos*) ducklings fed more than 30 mg As/kg diet as sodium arsenate	17
Small Laboratory Mammals		
Adverse effects, sensitive species	Single oral dose 2.5–33.0 mg As/kg BW; chronic doses 1–10 mg As/kg BW; 50 mg As/kg diet	7
Domestic Livestock		
Feedstuffs	Usually <2 mg total As/kg FW; <4 mg total As/kg in grasses and <10 in fish meals	18
Tissue residues		
Normal, muscle	<0.3 mg total As/kg FW	19
Arsenic-poisoned, liver and kidney	5–10 mg total As/kg FW	18, 20

[a]References: 1, USPHS 2000; 2, Jelinek and Corneliussen 1977; 3, Jewett and Naidu 2000; 4, Adams et al. 1993; 5, Kurttio et al. 1998; 6, Huang and Dasgupta 1999; 7, Eisler 2000; 8, Matschullat et al. 2000; 9, NRCC 1978; 10, NAS 1977; 11, Pershagen and Vahter 1979; 12, O'Neill 1990; 13, Sorensen et al. 1985; 14, USEPA 1985; 15, Goede 1985; 16, Custer et al. 2002; 17, Camardese et al. 1990; 18, Vreman et al. 1986; 19, van der Veen and Vreman 1986; 20, Thatcher et al. 1985.

µg/L, and behavioral impairment of goldfish (*Carassius auratus*) at 100 µg/L. A downward adjustment in the current freshwater aquatic life protection criterion seems merited. A similar scenario exists for saltwater life protection, where the water quality criterion of 36 µg As^{3+}/L was reduced from 508 µg As^{3+}/L 5 yr earlier (USEPA 1980, 1985), with only a few species of algae showing adverse effects at <36 µg As/L (e.g., reduced growth at 19–22 µg/L).

Arsenic criteria in marine products of commerce also need to be reexamined because most of the arsenic in seafoods is in the form of arsenobetaine or some other comparatively harmless form and does not pose a threat to the consumer. It is now clear that the formulation of maximum permissible concentrations of arsenic in seafoods for health regulation purposes should recognize the chemical nature of arsenic (Jelinek and Corneliussen 1977; Phillips et al. 1982; Ozretic et al. 1990; McGeachy and Dixon 1990; Eisler 2000).

Various phenylarsonic acids, including arsanilic acid, sodium arsinilate, and 3-nitro-4-hydroxyphenylarsonic acid, have been used as feed additives for disease control and for improvement of weight gain in swine and poultry for more than 40 yr (NAS 1977). The arsenic is present as As^{5+} and is rapidly excreted; present regulations require withdrawal of arsenical feed additives 5 d before slaughter for satisfactory feed depuration (NAS 1977). Under these conditions, total arsenic residues in edible tissues do not exceed the maximum permissible limit of 3 mg/kg FW (Jelinek and Corneliussen 1977). Organoarsenicals will probably continue to be used as feed additives until new evidence indicates the contrary.

Many authorities now recognize that current arsenic criteria are not sufficient for adequate protection and that additional data are required for meaningful arsenic standards [NAS 1977; USEPA 1980, 1985; Abernathy et al. 1997; Society for Environmental Geochemistry and Health (SEGH) 1998; Eisler 2000]. Specifically, there is general agreement that data are needed on the following subjects:

1. Cancer incidence and other abnormalities in natural resources with elevated arsenic levels, and the relation to potential carcinogenicity of arsenic compounds.
2. Interaction effects of arsenic with other carcinogens, cocarcinogens, promoting agents, inhibitors, and common environmental contaminants.
3. Controlled studies with aquatic and terrestrial indicator organisms on physiological and biochemical effects of long-term, low-dose exposures to inorganic and organic arsenicals, including effects on reproduction and genetic makeup.
4. Methodologies for establishing maximum permissible tissue concentrations for arsenic.
5. Effects of arsenic in combination with infectious agents.
6. Mechanisms of arsenical growth-promoting agents.
7. Role of arsenic in nutrition.
8. Extent of animal adaptation to arsenicals and the mechanisms of action.
9. Identification and quantification of mineral and chemical forms of arsenic in rocks, soils, and sediments that constitute the natural forms of arsenic entering water and the food chain.
10. Physicochemical processes influencing arsenic cycling.

Summary

Arsenic sources to the biosphere associated with gold mining include waste soil and rocks, residual water from ore concentrations, roasting of some types of gold-containing ores to remove sulfur and sulfur oxides, and bacterially enhanced leaching. Arsenic concentrations near gold mining operations are elevated in abiotic materials and biota: maximum total arsenic concentrations measured were 560 µg/L in surface waters, 5.16 mg/L in sediment pore waters, 5.6 mg/kg DW in bird liver, 27 mg/kg DW in terrestrial grasses, 50 mg/kg DW in soils, 79 mg/kg DW in aquatic plants, 103 mg/kg DW in bird diets, 225 mg/kg DW in soft parts of bivalve molluscs, 324 mg/L in mine drainage waters, 625 mg/kg DW in aquatic insects, 7,700 mg/kg DW in sediments, and 21,000 mg/kg DW in tailings.

Single oral doses of arsenicals that were fatal to 50% of tested species ranged from 17 to 48 mg/kg BW in birds and from 2.5 to 33 mg/kg BW in mammals. Susceptible species of mammals were adversely affected at chronic doses of

1–10 mg As/kg BW or 50 mg As/kg diet. Sensitive aquatic species were damaged at water concentrations of 19–48 µg As/L, 120 mg As/kg diet, or tissue residues (in the case of freshwater fish) >1.3 mg/kg fresh weight. Adverse effects to crops and vegetation were recorded at 3–28 mg of water-soluble As/L (equivalent to about 25–85 mg total As/kg soil) and at atmospheric concentrations >3.9 µg As/m^3. Gold miners had a number of arsenic-associated health problems, including excess mortality from cancer of the lung, stomach, and respiratory tract. Miners and schoolchildren in the vicinity of gold mining activities had elevated urine arsenic of 25.7 µg/L (range, 2.2–106.0 µg/L). Of the total population at this location, 20% showed elevated urine arsenic concentrations associated with future adverse health effects; arsenic-contaminated drinking water is the probable causative factor of elevated arsenic in their urine. Proposed arsenic criteria to protect human health and natural resources are listed and discussed. Many of these proposed criteria do not adequately protect sensitive species.

Acknowledgments

I thank Wanda Manning for library services and Thomas W. Custer and Glenn H. Olsen for their insightful comments on an early draft.

References

Abernathy CO, Calderon RL, Chappell WR (eds) (1997) Arsenic. Exposure and Health Effects. Chapman & Hall, London.

Adams MA, Bolger M, Carrington CD, Coker CE, Cramer GM, DiNovi MJ, Dolan S (1993) Guidance document for arsenic in shellfish. U.S. Food and Drug Admin, Washington, DC.

Adams MD, Johns MW, Dew DW (1999) Recovery of gold from ores and environmental aspects. In: Schmidbaur H (ed) Gold: Progress in Chemistry, Biochemistry and Technology. Wiley, New York, pp 66–104.

Agate AD (1996) Recent advances in microbial mining. World J Microbiol Biotechnol 12:487–495.

Amonoo-Neizer EH, Amekor EMK (1993) Determination of total arsenic in environmental samples from Kumasi and Obuasi, Ghana. Environ Health Perspect 101:46–49.

Amonoo-Neizer EH, Nyamah D, Bakiamoh SB (1996) Mercury and arsenic pollution in soil and biological samples around the mining town of Obuasi, Ghana. Water Air Soil Pollut 91:363–373.

Asperger S, Cetina-Cizmek B (1999) Metal complexes in tumour therapy. Acta Pharm 49:225–236.

Azcue JM, Mudroch A, Rosa F, Hall GEM (1994) Effects of abandoned gold mine tailings on the arsenic concentrations in water and sediments of Jack of Clubs Lake, B.C. Environ Technol 15:669–678.

Boffetta P, Kogevinas M, Pearce N, Matos E (1994) Cancer. In: Occupational Cancer in Developing Countries. IARC Sci Publ 129. Oxford University Press, New York, pp 111–126.

Bowell RJ, Warren A, Minjera HA, Kimaro N (1995) Environmental impact of former gold mining on the Orangi River, Serengeti N.P., Tanzania. Biogeochemistry 28: 131–160.

Bright DA, Coedy B, Dushenko WT, Reimer KJ (1994) Arsenic transport in a watershed receiving gold mine effluent near Yellowknife, Northwest Territories, Canada. Sci Total Environ 155:237–252.

Bright DA, Dodd M, Reimer KJ (1996) Arsenic in sub-Arctic lakes influenced by gold mine effluent: the occurrence of organoarsenicals and 'hidden' arsenic. Sci Total Environ 180:165–182.

Cain DJ, Luoma SN, Carter JL, Fend SV (1992) Aquatic insects as bioindicators of trace element contamination in cobble-bottom rivers and streams. Can J Fish Aquat Sci 49: 2141–2154.

Camardese MB, Hoffman DJ, LeCaptain LJ, Pendleton GW (1990) Effects of arsenate on growth and physiology in mallard ducklings. Environ Toxicol Chem 9:785–795.

Cockell KA, Hilton JW (1985) Chronic toxicity of dietary inorganic and organic arsenicals to rainbow trout (*Salmo gairdneri* R.). Fed Proc 44(4):938.

Custer TW, Custer CM, Larson S, Dickerson KK (2002) Arsenic concentrations in house wrens from Whitewood Creek, South Dakota, USA. Bull Environ Contam Toxicol 68:517–524.

Da Rosa CD, Lyon JS (eds) (1997) Golden dreams, poisoned streams. Mineral Policy Center, Washington, DC.

Deknudt G, Leonard A, Arany J, Du Buisson GJ, Delavignette E (1986) *In vivo* studies in male mice on the mutagenic effects of inorganic arsenic. Mutagenesis 1:33–34.

Eisler R (2000) Arsenic. In: Handbook of Chemical Risk Assessment: Health Hazards to Humans, Plants, and Animals, vol 3. Lewis, Boca Raton, pp 1501–1566.

Galbraith H, LeJeune K, Lipton J (1995) Metal and arsenic impacts to soils, vegetation communities and wildlife habitat in southwest Montana uplands contaminated by smelter emissions. I. Field evaluation. Environ Toxicol Chem 14:1895–1903.

Goede AA (1985) Mercury, selenium, arsenic and zinc in waders from the Dutch Wadden Sea. Environ Pollut 37A:287–309.

Golow AA, Schleuter A, Amihere-Mensah S, Granson HLK, Tetteh MS (1996) Distribution of arsenic and sulphate in the vicinity of Ashanti goldmine at Obuasi, Ghana. Bull Environ Contam Toxicol 56:703–710.

Greer J (1993) The price of gold: environmental costs of the new gold rush. Ecologist 23(3):91–96.

Grosser JR, Hagelgans V, Hentschel T, Priester M (1994) Heavy metals in stream sediments: a gold mining area near Los Andes, southern Columbia S.A. Ambio 23:146–149.

Hallberg KB, Sehlin HM, Lindstrom EB (1996) Toxicity of arsenic during high temperature bioleaching of gold-bearing arsenical pyrite. Appl Microbiol Biotechnol 45:212–216.

Hamasaki T, Nagase H, Yoshioka Y, Sato T (1995) Formation, distribution, and ecotoxicity of methylmetals of tin, mercury, and arsenic in the environment. Crit Rev Environ Sci Technol 25:45–91.

Hoffman DJ, Sanderson CJ, LeCaptain LJ, Cromartie E, Pendleton GW (1992) Interactive effects of arsenate, selenium, and dietary protein on survival, growth, and physiology in mallard ducklings. Arch Environ Contam Toxicol 22:55–62.

Holson JF, DeSesso JM, Scialli AR, Farr CF (1998) Inorganic arsenic and prenatal development: a comprehensive evaluation for human risk assessment. In: Society for Environmental Geochemistry and Health (SEGH) 3rd International Conference on Arsenic Exposure Health Effects, San Diego, California, July 12–15, 1998, p 23.

Hood RD (1985) Cacodylic acid: agricultural uses, biologic effects, and environmental fate. VA Monograph. Available from Supt. Documents, US Govt Printing Office, Washington, DC.

Hood RD, Vedel-Macrender GC, Zaworotko MJ, Tatum FM, Meeks RG (1987) Distribution, metabolism, and fetal uptake of pentavalent arsenic in pregnant mice following oral or intraperitoneal administration. Teratology 35:19–25.

Hopenhayn-Rich C, Johnson KD, Hertz-Picciotto J (1998) Reproductive and developmental effects associated with chronic arsenic exposure. In: Society for Environmental Geochemistry and Health (SEGH), 3rd International Conference on Arsenic Exposure Health Effects, San Diego California, July 12–15, 1998, p. 21.

Huang H, Dasgupta PK (1999) A field-deployable instrument for the measurement and speciation of arsenic in potable water. Anal Chim Acta 380:27–37.

Hudson RH, Tucker RK, Haegle MA (1984) Handbook of Toxicity of Pesticides to Wildlife. Resource Publ 153. U.S. Fish and Wildlife Service, Washington, DC.

Hughes MF, Kenyon EM (1998) Dose-dependent effects on the disposition of monomethylarsonic acid and dimethylarsinic acid in the mouse after intravenous administration. J Toxicol Environ Health 53A:95–112.

Jauge P, Del-Razo LM (1985) Uric acid levels in plasma and urine in rats chronically exposed to inorganic As(III) and As(V). Toxicol Lett 26:31–35.

Jelinek CF, Corneliussen PE (1977) Levels of arsenic in the United States food supply. Environ Health Perspect 19:83–87.

Jenkins DW (1980) Biological Monitoring of Toxic Trace Metals, vol 2. Toxic Trace Metals in Plants and Animals of the World. Part 1. Report 600/3-80-090. U.S. Environmental Protection Agency, Washington, DC, pp 30–138.

Jewett SC, Naidu S (2000) Assessment of heavy metals in red king crabs following offshore placer gold mining. Mar Pollut Bull 40:478–490.

Johnson WW, Finley MT (1980) Handbook of Acute Toxicity of Chemicals to Fish and Aquatic Invertebrates. Resource Publ 137. U.S. Fish and Wildlife Service, Washington, DC.

Johnson SW, Rice SD, Moles DA (1998a) Effects of submarine mine tailings disposal on juvenile yellowfin sole (*Pleuronectes asper*): a laboratory study. Mar Pollut Bull 36:278–287.

Johnson SW, Stone RP, Love DC (1998b) Avoidance behavior of ovigerous tanner crabs *Chionoecetes bairdi* exposed to mine tailings: a laboratory study. Alaska Fish Res Bull 5:39–45.

Kabir H, Bilgi C (1993) Ontario gold miners with lung cancer. J Occup Med 35:1203–1207.

Kurttio P, Komulainen H, Hakala E, Pekkanen J (1998) Urinary excretion of arsenic species after exposure to arsenic present in drinking water. Arch Environ Contam Toxicol 34:297–305.

Kusiak RA, Ritchie AC, Springer J, Muller J (1993) Mortality from stomach cancer in Ontario miners. Br J Ind Med 50:117–126.

Lau S, Mohamed M, Yen ATC, Su'ut S (1998) Accumulation of heavy metals in freshwater molluscs. Sci Total Environ 214:113–121.

Leduc G (1984) Cyanides in water: toxicological significance. In: Weber JL (ed) Aquatic Toxicology, vol 2. Raven Press, New York, pp 153–224.

Lima AR, Curtis C, Hammermeister DE, Markee TP, Northcutt CE, Brooke LT (1984)

Acute and chronic toxicities of arsenic (III) to fathead minnows, flagfish, daphnids, and an amphipod. Arch Environ Contam Toxicol 13:595–601.

Marques IA, Anderson LE (1986) Effects of arsenite, sulfite, and sulfate on photosynthetic carbon metabolism in isolated pea (*Pisum sativum* L., cv Little Marvel) chloroplasts. Plant Physiol 82:488–493.

Matschullat J, Borba RP, Deschamps E, Figueiredo BR, Gabrio T, Schwenk M (2000) Human and environmental contamination in the iron quadrangle, Brazil. Appl Geochem 15:181–190.

May TW, Wiedmeyer RH, Gober J, Larson S (2001) Influence of mining-related activities on concentrations of metals in water and sediment from streams of the Black Hills, South Dakota. Arch Environ Contam Toxicol 40:1–9.

McGeachy SM, Dixon DG (1990) Effect of temperature on the chronic toxicity of arsenate to rainbow trout (*Oncorhynchus mykiss*). Can J Fish Aquat Sci 47:2228–2234.

Meharg AA, Shore RF, Broadgate KF (1998) Edaphic factors affecting the toxicity and accumulation of arsenate in the earthworm *Lumbricus terrestris*. Environ Toxicol Chem 17:1124–1131.

Nagymajtenyi L, Selypes A, Berencsi G (1985) Chromosomal aberrations and fetotoxic effects of atmospheric arsenic exposure in mice. J Appl Toxicol 5:61–63.

Naqvi SM, Flagge CT (1990) Chronic effects of arsenic on American red crayfish, *Procambarus clarki*, exposed to monosodium methanearsonate (MSMA) herbicide. Bull Environ Contam Toxicol 45:101–106.

National Academy of Sciences (NAS) (1977) Arsenic. NAS, Washington, DC.

National Research Council of Canada (NRCC) (1978) Effects of arsenic in the Canadian environment. Publ 15391. NRCC, available from publications NRCC/CNRC, Ottawa, Ontario, Canada

Ng JC, Seawright AA, Qi L, Garnett CM, Moore MR, Chiswell B (1998) Tumours in mice induced by chronic exposure of high arsenic concentrations in drinking water. In: Society for Environmental Geochemistry and Health (SEGH) 3rd International Conference on Arsenic Exposure Health Effects, San Diego, California, July 12–15, 1998, p 28.

Ngubane WT, Baecker AAW (1990) Oxidation of gold-bearing pyrite and arsenopyrite by *Sulfolobus acidocaldarius* and *Sulfolobus* BC in airlift reactors. Biorecovery 1: 255–259.

Nystrom RR (1984) Cytological changes occurring in the liver of coturnix quail with an acute arsenic exposure. Drug Chem Toxicol 7:587–594.

O'Neill P (1990) Arsenic. In: Heavy Metals in Soils. Halsted Press, Glasgow, pp 83–99.

Ozretic B, Krajinovic-Ozretic M, Santin J, Medjugorac B, Kras M (1990) As, Cd, Pd, and Hg in benthic animals from the Kvarber-Rijeka region, Yugoslavia. Mar Pollut Bull 21:595–597.

Pain S (1987) After the goldrush. New Sci 115(1574):36–40.

Passino DRM, Novak AJ (1984) Toxicity of arsenate and DDT to the cladoceran *Bosmina longirostris*. Bull Environ Contam Toxicol 33:325–329.

Pendleton GW, Whitworth MR, Olsen GH (1995) Accumulation and loss of arsenic and boron, alone and in combination, in mallard ducks. Environ Toxicol Chem 14: 1357–1364.

Pershagen G, Vahter M (1979) Arsenic—a Toxicological and Epidemiological Appraisal. Naturvards-verket Rapp SNV PM 1128. Liber Tryck, Stockholm.

Phillips DJH, Thompson GB, Gabuji KM, Ho CT (1982). Trace metals of toxicological significance to man in Hong Kong seafood. Environ Pollut 3B:27–45.

Rahn PH, Davis AD, Webb CJ, Nichols AD (1996) Water quality impacts from mining in the Black Hills, South Dakota, USA. Environ Geol 27:38–53.

Ripley EA, Redmann RE, Crowder AA (1996) Environmental effects of mining. St. Lucie Press, Delray Beach, FL.

Robertson ID, Harms WE, Ketterer PJ (1984) Accidental arsenic toxicity of cattle. Aust Vet J 61:366–367.

Robertson JL, McLean JA (1985) Correspondence of the LC_{50} for arsenic trioxide in a diet-incorporation experiment with the quantity of arsenic ingested as measured by X-ray, energy-dispersive spectrometry. J Econ Entomol 78:1035–1036.

Samecka-Cymerman A, Kempers AJ (1998) Bioindication of gold by aquatic bryophytes. Acta Hydrochim Hydrobiol 26:90–94.

Sanders JG (1986) Direct and indirect effects of arsenic on the survival and fecundity of estuarine zooplankton. Can J Fish Aquat Sci 43:694–699.

Santini JM, Sly LI, Schnagl RD, Macy JM (2000) A new chemolithoautotrophic arsenite-oxidizing bacterium isolated from a gold mine: phylogenetic, physiological and preliminary biochemical studies. Appl Environ Microbiol 66:92–97.

Savabieasfahani M, Lochmiller RL, Rafferty DP, Sinclaiar JA (1998) Sensitivity of wild cotton rats (*Sigmodon hispidus*) to the immunotoxic effects of low-level arsenic exposure. Arch Environ Contam Toxicol 34:289–296.

Selby LA, Case AA, Osweiler GD, Hages HM Jr (1977) Epidemiology and toxicology of arsenic poisoning in domestic animals. Environ Health Perspect 19:183–189.

Sheppard MI, Thibault DH, Sheppard SC (1985) Concentrations and concentration ratios of U, As and Co in Scots pine grown in a waste-site soil and an experimentally contaminated soil. Water Air Soil Pollut 26:85–94.

Simonato L, Moulin JJ, Javelaud B, Ferro G, Wild P, Winkelmann R, Saracci R (1994) A retrospective mortality study of workers exposed to arsenic in a gold mine and refinery in France. Am J Ind Med 25:625–633.

Society for Environmental Geochemistry and Health (SEGH) (1998) Abstracts, Third International Conference on Arsenic Exposure and Health Effects, San Diego, CA, July 12–15, 1998.

Sorensen EMB, Mitchell RR, Pradzynski A, Bayer TL, Wenz LL (1985) Stereological analyses of hepatocyte changes parallel arsenic accumulation in the livers of green sunfish. J Environ Pathol Toxicol Oncol 6:195–210.

Spehar RL, Fiandt JT, Anderson RL, DeFoe DL (1980) Comparative toxicity of arsenic compounds and their accumulation in invertebrates and fish. Arch Environ Contam Toxicol 9:53–63.

Stanley TR, Spann JW, Smith GJ, Rosscoe R (1994) Main and interactive effects of arsenic and selenium on mallard reproduction and duckling growth and survival. Arch Environ Contam Toxicol 26:444–451.

Stine ER, Hsu CA, Hoovers TD, Aposhian HV, Carter DE (1984) *N*-(2,3-Dimercaptopropyl)phthalamidic acid: protection *in vivo* and *in vitro* against arsenic intoxication. Toxicol Appl Pharmacol 75:329–336.

Stone RP, Johnson SW (1997) Survival, growth, and bioaccumulation of heavy metals by juvenile tanner crabs (*Chionoecetes bairdi*) held on weathered mine tailings. Bull Environ Contam Toxicol 58:830–837.

Stone RP, Johnson SW (1998) Prolonged exposure to mine tailings and survival and reproductive success of ovigerous tanner crabs (*Chionoecetes bairdi*). Bull Environ Contam Toxicol 61:548–556.

Tarras-Wahlberg NH, Flachier A, Fredriksson G, Lane S, Lundberg B, Sangfors O

(2000) Environmental impact of small-scale and artisanal gold mining in southern Ecuador. Ambio 29:484–491.

Thatcher CD, Meldrum JB, Wikse SE, Whittier WD (1985) Arsenic toxicosis and suspected chromium toxicosis in a herd of cattle. J Am Vet Assoc 187:179–182.

Thursby GB, Steele RL (1984) Toxicity of arsenite and arsenate to the marine macroalgae *Champia parvula* (Rhodophyta). Environ Toxicol Chem 52:641–648.

U.S. Bureau of Land Management (USBLM) (2000) Cumulative impact analysis of dewatering and water management operations for the Betze Project, South Operations Area Amendment, and Leevile Project. USBLM, Elko, NV.

U.S. Environmental Protection Agency (USEPA) (1980) Ambient water quality criteria for arsenic. Report 440/5-80-021. Washington, DC.

U.S. Environmental Protection Agency (USEPA) (1985) Ambient water quality criteria for arsenic—1984. Report 440/5-84-033. EPA, Washington, DC.

U.S. Public Health Service (USPHS) (2000) Toxicological profile for arsenic (update). Draft for public comment. Agency Toxic Substances Diseases Registry. USPHS, Washington, DC.

van der Veen NG, Vreman K (1986) Transfer of cadmium, lead, mercury and arsenic from feed into various organs and tissues of fattening lambs. Neth J Agric Sci 34: 134–153.

Vreman K, van der Veen NG, van der Molen EJ, de Ruig WG (1986) Transfer of cadmium, lead, mercury and arsenic from feed into milk and various tissues of dairy cows: chemical and pathological data. Neth J Agric Sci 34:129–144.

Wang DS, Weaver RW, Melton JR (1984) Microbial decomposition of plant tissue contaminated with arsenic and mercury. Environ Pollut 34A:275–282.

Webb DR, Wilson SE, Carter DE (1986) Comparative pulmonary toxicity of gallium arsenide, gallium (III) oxide or arsenic (III) oxide intratracheally instilled into rats. Toxicol Appl Pharmacol 82:405–416.

Winski SL, Carter DE (1998) Arsenate toxicity in human erythrocytes: characterization of morphologic changes and determination of the mechanism of damage. J Toxicol Environ Health 53A:345–355.

Wong HKT, Gauthier A, Nriagu JO (1999) Dispersion and toxicity of metals from abandoned gold mine tailings at Goldenville, Nova Scotia, Canada. Sci Total Environ 228: 35–47.

Woolson EA (ed) (1975) Arsenical Pesticides. ACS Symposia, Services 7. American Chemical Society, Washington, DC.

Yoshida K, Inoue Y, Kuroda K, Chen H, Wanibuchi H, Fukushima S, Endo G (1998) Urinary excretion of arsenic metabolites after long-term oral administration of various arsenic compounds to rats. J Toxicol Environ Health 54A:179–192.

Manuscript received February 1; accepted March 17, 2003.

Cumulative and Comprehensive Subject Matter Index Volumes 171–180

Abamectin, avermectin B$_1$, **171**:112
Abamectin, effects on crop pests/beneficial insects, **171**:115
Abamectin, environmental effects, **171**:115
Abamectin, non-target insect effects, **171**:116
Abamectin, resistance induction insect pests, **171**:116
Abbreviations, scientific organizations, **172**:119
Abbreviations, toxicological, **176**:4
Abiotic coupling, manganese oxidation of Cr(III), **178**:121
Abiotic hydrolysis, pesticides aquatic environment, **175**:79 ff.
Abiotic pesticide hydrolysis, adsorption effects, **175**:85
Abiotic pesticide hydrolysis, clay effects, **175**:88
Abiotic pesticide hydrolysis, cosolvent effects, **175**:85
Abiotic pesticide hydrolysis, dissolved organic matter effects, **175**:86
Abiotic pesticide hydrolysis, metal ions/oxides effects, **175**:88
Abiotic pesticide hydrolysis, micelle effects, **175**:89
Abiotic pesticide hydrolysis, modifying environmental factors, **175**:83
Abiotic pesticide hydrolysis, pH rate profiles, **175**:82
Absorption coefficients (photochemical), organophosphates (table), **172**:149
Acanthamoeba, bacterial endosymbionts, **180**:98
Acanthamoeba, currently identified species (table), **180**:95
Acanthamoeba eye infections, dose response, **180**:17
Acanthamoeba, health effects, **180**:103

Acanthamoeba, human infections caused, **180**:116
Acanthamoeba, in air, dust & soil, **180**:103
Acanthamoeba, in animal wastes, **180**:102
Acanthamoeba, in sewage & biosolids, **180**:102
Acanthamoeba, in surface waters, **180**:100
Acanthamoeba, in swimming pools, spas, **180**:102
Acanthamoeba, in tapwater & bottled water, **180**:101
Acanthamoeba infection in AIDS patients, **180**:114
Acanthamoeba, interactions with *Legionella pneumophila*, **180**:114
Acanthamoeba keratitis, contact lenses association, **180**:119
Acanthamoeba keratitis, diagnosis, **180**:106
Acanthamoeba keratitis, eye infections, **180**:104
Acanthamoeba keratitis, immunity, **180**:110
Acanthamoeba keratitis, incidence & pathogenicity, **180**:109
Acanthamoeba keratitis, mechanisms involved in, **180**:118
Acanthamoeba keratitis, risk factors, **180**:121
Acanthamoeba keratitis, symptoms in patients, **180**:105
Acanthamoeba keratitis, treatment, **180**:108
Acanthamoeba, life cycle (diag.), **180**:97
Acanthamoeba, methods of identification, **180**:96
Acanthamoeba, previously genus *Hartmanella*, **180**:95
Acanthamoeba resistance to water treatments, **180**:117

Acanthamoeba species group classification, **180**:96
Acanthamoeba spp. health effects, **180**:93 ff.
Acanthamoeba spp. potential waterborne transmission, **180**:93 ff.
Acanthamoeba, where found, **180**: 99, 100
Acenaphthylene, PAH, **179**:75
Acenapthene, PAH, **179**:75
Acephate, hydrolysis pathways, **175**:140
Acephate, sublethal AChE inhibition, aquatic organisms, **172**:45
Acetylcholine (ACh), synapse neurotransmitter, **172**:150
Acetylcholinesterase (AChE), function in nervous system, **172**:22
Acetylcholinesterase (AChE) inhibition, organophosphates, **172**:2
Acetylcholinesterase (AChE), insecticidal activity role, **172**:150
Acetylcholinesterase inhibitors, tissue residues, **173**:17
AChE (acetylcholinesterase), function in nervous system, **172**:22
AChE (acetylcholinesterase) inhibition, organophosphates, **172**:2
AChE (acetylcholinesterase), insecticidal activity role, **172**:150
AChE inhibition, by non-OP or non-carbamate pesticides, **172**:52
AChE inhibition, organophosphate insecticidal mode of action, **172**:150
AChE inhibition, sublethal OP levels aquatic organisms, **172**:44
AChE inhibitors, CBRs, **173**:5
AChE inhibitors, mode of action, **173**:8
AChE inhibitors, tissue residues, **173**:17
Acid equivalent (a.e.), defined, **174**:20
Acridine & its homocyclic analog anthracene (illus.), **173**:41
Acridine, anaerobic degradation pathway, **173**:45
Acridine, photoenhanced toxicity different species (table), **173**:60
Acridine toxicity, species groups, **173**:50
Acrolein, hydrolytic profile, **175**:232
Actinometry, light source intensity, **172**:142

Acute toxicity, fipronil, **176**:34
Acute toxicity, fipronil mammals, **176**:34
ADI, fipronil, **176**:46
Adonis®, fipronil proprietary name, **176**:6
Adsorption, abiotic pesticide hydrolysis, **175**:85
Adsorption coefficient (K_d), methyl bromide, **177**:68
Adsorption/desorption coefficients soil, pyrethroids, **174**:56
Advisory Committee on Crop Protection Chemistry, **177**:125
a.e. (acid equivalent) defined, **174**:20
Aerodynamic method, methyl bromide volatilization, **177**:85
Aflatoxin, ammoniation decontamination safety, **171**:157
Aflatoxin B_1, conversion to aflatoxin M_1 in cow's milk, **171**:157
Aflatoxin B_1, most potent of the four aflatoxins, **171**:140
Aflatoxin hazard reduction, using ammoniation, **171**:139 ff.
Aflatoxin M_1, conversion from aflatoxin B_1 in cow's milk, **171**:157
Aflatoxin M_1, residues in Arizona milk, **171**:159
Aflatoxin, reduction using ammonia-related procedures, **171**:167
Aflatoxin, trout feeding studies, **171**:163
Aflatoxin-ammonia reaction products, feed concentrations, **171**:155
Aflatoxin-ammonia reaction products, toxic potential, **171**:156, 160
Aflatoxin-ammonia reaction products, toxicity, **171**:153, 160
Aflatoxin-ammonia reaction products, toxicological properties, **171**:160
Aflatoxin-ammoniation, feeding studies of contaminated feeds, **171**:160
Aflatoxin-contaminated feeds, mutagenic potentials, **171**:167
Aflatoxin-contaminated peanut meal, rat feeding studies, **171**:163
Aflatoxin-related ammonia decontamination pathway (fig.), **171**:152
Aflatoxins, ammonia decontamination efficacy, **171**:149, 150

Aflatoxins, ammonia treatment of contaminated cottonseed, **171**:148
Aflatoxins, ammoniation chemistry effect, **171**:151
Aflatoxins, ammoniation decontamination methods, **171**:147
Aflatoxins, animal feed levels vs edible tissue levels, **171**:142, 143
Aflatoxins, Arizona cottonseed monitoring results, **171**:146
Aflatoxins, breakdown products from ammoniation (fig.), **171**:152
Aflatoxins, carcinogenicity, **171**:140
Aflatoxins chemistry, ammoniation effect, **171**:151
Aflatoxins, FDA action levels animal feeds, **171**:143
Aflatoxins, FDA action levels cow's milk, **171**:144
Aflatoxins, FDA feedstuffs surveillance results, **171**:146
Aflatoxins, feed blending to reduce levels, **171**:144
Aflatoxins, hepatotoxicity, **171**:139
Aflatoxins, human liver cancer risk, **171**:141
Aflatoxins, legal history/action levels, **171**:142
Aflatoxins, list of contaminated feedstuffs, **171**:146
Age effect, hair trace element contamination, **175**:52
Aging effect on metal availability in soils, **178**:1 ff.
Aging, increases in metal availability, soils, **178**:9
Aging metals, bioavailability to invertebrates, **178**:11
Aging metals, soil availability, **178**:2
Ah-receptor agonists, CBRs, **173**:5
Ah-receptor agonists, defined, **173**:2
Ah-receptor agonists, mode of action, **173**:8
Ah-receptor agonists, tissue residues, **173**:20
AIDS patients, *Acanthamoeba* effects, **180**:114
Air pollution, hair trace element contamination, **175**:61
Air, proposed arsenic criteria, **180**:156
Air sampling adsorbents, methyl bromide, **177**:56
Air sampling, methyl bromide, **177**:54, 55
Alachlor, degradation by photo-Fenton process, **177**:160
Alachlor, hydrolytic profile, **175**:194
Alcohol drinking, hair trace element contamination, **175**:55
Aldicarb, catalysis by metal ions, **175**:184
Aldicarb, hydrolysis pathways, **175**:116
Aldicarb, hydrolytic profile, **175**:200
Aldicarb sulfoxide, hydrolytic profile, **175**:200
Aldoxycarb, hydrolytic profile, **175**:200
Aldrin, hydrolytic profile, **175**:206
Alfalfa leafcutter bee, spinosad toxicity, **179**:42
Algae, chromium toxicity, **178**:128
Algae, environmental metals risk monitor, **178**:23 ff.
Algae test protocols for toxicity, **178**:26
Algal Assay Procedure Bottle Test (EPA), **178**:23
Algal culture techniques, toxicity tests, **178**:27
Algal metal toxicity, effects measurements, **178**:37
Algal metal toxicity, regulatory context, **178**:25
Algal metal toxicity testing, **178**:23 ff.
Algal metal toxicity testing, acclimation/adaptation laboratory, **178**:41
Algal metal toxicity testing, adsorption, **178**:39
Algal metal toxicity testing, inoculum rate, **178**:38
Algal metal toxicity testing, lab/field extrapolation, **178**:42
Algal metal toxicity testing, toxicity mechanisms, **178**:39
Algal metal toxicity testing, uptake, **178**:39
Algal metal toxicity testing, variables affecting results, **178**:43
Algal sensitivity, toxicants, interspecific differences, **178**:35
Algal sensitivity, toxicants, intraspecies variability, **178**:36

Algal species used in standard toxicity testing, **178**:34
Algal toxicity testing, test media, **178**:28
Aliphatic dithiophosphates, photochemistry, **172**:195
Aliphatic thiophosphates, photochemistry, **172**:156
Alkyl *N*-arylcarbamate hydrolysis, **175**:112
Allethrin, dimer formation in alkaline hydrolysis, **175**:94
Allethrin, hydrolytic profile, **175**:192
Alloxidim sodium, Beckmann rearrangement, **175**:154
Alumina aging in soils, **178**:4
Amide herbicides, structures & hydrolytic profiles, **175**:194
Amide hydrolysis mechanisms, **175**:100
Amide pesticide hydrolysis, kinetics mechanisms, **175**:99
Amide pesticides, Classes 1, 2, and 3 hydrolysis, **175**:101
Amidosulfuron, hydrolytic profile, **175**:224
Amino acids, chlorine effects, **171**:6
Aminoglycosides, mechanisms of action, **171**:39
Amitraz, hydrolytic profile, **175**:232
Ammonia, effect on feed composition/animal performance, **171**:157
Ammonia treatment of aflatoxin-contaminated cottonseed, **171**:148
Ammonia-aflatoxin reaction products, feed concentrations, **171**:155
Ammonia-aflatoxin reaction products, toxic potential, **171**:156, 160
Ammonia-aflatoxin reaction products, toxicity, **171**:153, 160
Ammonia-aflatoxin reaction products, toxicological properties, **171**:160
Ammonia-treated aflatoxin-contaminated feeds, mutagenic potentials, **171**:167
Ammoniated cottonseed, interstate shipment illegal, **171**:149
Ammoniated cottonseed, states permitting use, **171**:149
Ammoniation, aflatoxin decontamination safety, **171**:157
Ammoniation, aflatoxin hazard reduction, **171**:139 ff.
Ammoniation, aflatoxin-related breakdown products (fig.), **171**:152
Ammoniation, aflatoxins reduction methods, **171**:147
Ammoniation, corn aflatoxin decontamination efficacy, **171**:150
Ammoniation, effect on aflatoxin chemistry, **171**:151
Ammoniation, feeding studies of aflatoxin-contaminated feeds, **171**:160
Amphibians, arsenic effects, **180**:144
Ampicillin resistant bacteria, **171**:23
Anabaena flos-aquau (alga), toxicity testing, **178**:34
Anaerobic degradation pathway, acridine, **173**:45
Anaerobic packed-bed bioreactor, Cr(VI) water remediation, **178**:145
Analytical methods, hair trace element contamination, **175**:63, 65
Analytical methods, house dust contaminants, **175**:13
Analytical methods, pesticide trace amounts, **175**:80
Animal growth promoters, arsenic, **180**:2
Animal nutrition, essential elements defined, **177**:2
Anodic stripping, hair trace element analysis, **175**:65
Antagonism, fipronil insecticide, **176**:5
Antarctic coast, anthropogenic metal pollutant biomonitoring, **171**:80
Antarctic cold waters, salinity, **171**:58
Antarctic freshwater ecosystems, trace metals, **171**:93
Antarctic lichens, baseline trace metal levels, **171**:87, 92
Antarctic marine ecosystems, **171**:57
Antarctic mosses, elemental compositions, **171**:89, 92
Antarctic organisms, trace metals, **171**:53 ff.
Antarctic terrestrial ecosystems, trace metals, **171**:85
Antarctic terrestrial fauna, identified, **171**:86

Antarctic terrestrial flora, identified, **171:** 86

Antarctic trace metals in circumpolar biomonitoring, **171:**53 ff.

Anthelmintics, veterinary soil contaminants, **180:**29

Anthracene, PAH, **179:**75

Anthropogenic metal pollutants, biomonitoring Antarctic coast, **171:**80

Anthropogenic processes, heavy metals in soils, **177:**4

Antibacterials, toxicity *Daphnia magna*, **180:**71

Antibacterials, veterinary soil contaminanats, **180:**29

Antibiotic/chlorine interaction, resistant bacteria, **171:**21

Antibiotic resistance, bacteria, **171:**19

Antibiotic resistance in bacteria, **171:**1 ff.

Antibiotic resistance in bacteria, UV use, **171:**34

Antibiotic-resistant coliforms, sewage effluent, **171:**23, 24, 36

Antibiotic-resistant fecal coliforms, sewage plant, **171:**25, 37

Antibiotics, mechanisms of action in bacteria, **171:**39

Antimony, house dust concentrations, **175:**30

Apatite, adsorption of heavy metals, **177:**21

APHA, American Public Health Association, **178:**34

Aquaculture medicines, environmental pathways, **180:**13

Aquaculture medicines, veterinary, **180:** 12

Aquatic biota, proposed arsenic criteria, **180:**157

Aquatic ecosystems, contaminated sediment effects, **174:**4

Aquatic environment, abiotic pesticide hydrolysis, **175:**79 ff.

Aquatic half-lives, pesticides, **175:** 190–233

Aquatic half-lives, pyrethroids, **174:**52

Aquatic invertebrates, bioindicators of environmental pollution, **172:**23

Aquatic invertebrates, ChE biomarker OP exposure, **172:**23

Aquatic organisms, avermectin effects, **171:**125

Aquatic organisms, CBRs, **173:**2

Aquatic organisms, contaminant residues, **173:**1 ff.

Aquatic organisms, exposure vs toxicological effects, **173:**4

Aquatic toxicity, veterinary medicines, **180:**44 ff.

Aquatic toxicity, veterinary pesticides, **180:**44 ff.

AQUIRE database, website address, **173:** 48

Archeaobacteria, dominant in extreme environments, **173:**137

Aromatic phosphates, organophosphate photochemistry, **172:**152

Aromatic thiophosphates, photochemistry, **172:**156

Arsenic contamination, phytoremediation gold site, **180:**141

Arsenic, effects amphibians, **180:**144

Arsenic, effects birds, **180:**146

Arsenic, effects domestic animals, **180:** 143

Arsenic, effects domestic animals, **180:** 148

Arsenic, effects freshwater fishes, **180:** 144

Arsenic, effects freshwater invertebrates, **180:**143

Arsenic, effects human health, **180:**152

Arsenic, effects mammals, **180:**148

Arsenic, effects marine fish, **180:**145

Arsenic effects, sensitive species, **180:**142

Arsenic, effects terrestrial invertebrates, **180:**146

Arsenic, effects terrestrial plants, **180:**145

Arsenic, environmental contamination, **180:**133 ff.

Arsenic, feed additives for livestock, **180:** 158

Arsenic, gold mining sources, **180:**133

Arsenic, groundwater contamination, gold mining, **180:** 134

Arsenic hazards to humans, **180:**133 ff.

Arsenic, health effects humans, **180**:143
Arsenic, historical human disease treatments, **180**:135
Arsenic, house dust concentrations, **175**:29
Arsenic, human diseases caused, **180**:136
Arsenic, in biota near gold extraction, **180**:137
Arsenic, lethal/sublethal effects various plants, **180**:143
Arsenic levels, biota & abiota near gold mining (table), **180**:138
Arsenic levels, international gold mining, **180**:138
Arsenic, occurrence in gold ore, **180**:133
Arsenic poisoning, acute episodes described, **180**:154
Arsenic poisoning, mammals (detailed table), **180**:154
Arsenic poisoning, subacute episodes, **180**:154
Arsenic, proposed criteria environmental protection, **180**:156
Arsenic, proposed criteria human protection, **180**:156
Arsenic, risks to human health, **180**:135
Arsenic, soil contaminant sources, **177**:3
Arsenic tailings, fish avoidance, **180**:153
Arsenic trioxide, atmospheric from gold ore roasting, **180**:134
Arsenic, various valences from gold mining, **180**:134
Arsenopyrite, arsenic source in gold ore, **180**:133
Aryl *N*-alkylcarbamate hydrolysis, **175**:113
Aryloxyalkanoate herbicides, hydrolysis, **175**:91
Aryloxyalkanoate herbicides, structures & hydrolytic profiles, **175**:190
Asana®, esfenvalerate, **174**:117
Aspergillus flavus, source of aflatoxins, **171**:139
Aspergillus parasiticus, source of aflatoxins, **171**:139
ASTM, American Society for Testing and Materials, **178**:34
Atmospheric arsenic trioxide, gold ore roasting, **180**:134

Atomic absorption spectrometry, hair trace element analysis, **175**:65
Atomic absorption spectrometry, house dust contaminant analysis, **175**:14
ATP formation, chlorine effects, **171**:8
Atrazine, hydrolytic profile, **175**:230
Atropisomer of methylsulfonyl-PCBs, **173**:91
Atropisomers of PCBs, chiral PCBs, **173**:91
Avermectin B_1, known as abamectin, **171**:112
Avermectin complex, four major components, **171**:112
Avermectin, earthworm effects, **171**:121, 123
Avermectin, environmental breakdown/fate, **171**:113
Avermectin, physicochemical properties, **171**:113
Avermectin, soil-inhabiting invertebrate effects, **171**:120
Avermectins, aquatic organisms effects, **171**:125
Avermectins, bird toxicity, **171**:126
Avermectins, chemistry and fate, **171**:112
Avermectins, effects on dung/organic matter breakdown, **171**:122
Avermectins, effects on plants, **171**:125
Avermectins, environmental impact, **171**:111 ff..
Avermectins, environmental risk assessment, **171**:127
Avermectins, mammalian toxicity, **171**:126
Avermectins, mode of action, **171**:112
Avermectins, sediment-inhabiting organisms effects, **171**:125
Avermectins, soil antimicrobial effects, **171**:124
Avermectins, uses, **171**:112
Avermectins, vertebrate toxicity, **171**:126
Avermectins, wildlife effects, **171**:126
Azaarene toxicity, **173**:39 ff.
Azaarenes, biotransformation and toxicity, **173**:52
Azaarenes, carcinogenicity, **173**:62
Azaarenes, chronic effect different species (graph), **173**:54

Azaarenes, chronic toxicity, **173**:52
Azaarenes, defined, **173**:40, 42
Azaarenes, direct toxicity, **173**:48
Azaarenes, genotoxicity, **173**:62
Azaarenes, K_{ow} related to baseline toxicity, **173**:48
Azaarenes, metabolites in the environment, **173**:47
Azaarenes, photo-products affect photosynthesis of marine diatom, **173**:61
Azaarenes, photochemical reactions, mechanisms & kinetics, **173**:53
Azaarenes, photochemical transformation, **173**:53
Azaarenes, photoenhanced toxicity, **173**:59
Azaarenes, phototoxic effects, **173**:53
Azaarenes, risk assessment, **173**:69
Azaarenes, teratogenicity, **173**:66
Azaarenes, toxic effects several to different species (graph), **173**:51
Azaarenes vs homocyclic PAHs, comparative toxicity, **173**:67
Azaarenes with more than three aromatic rings, microbial degradation, **173**:46
Azimsulfuron, hydrolytic profile, **175**:224
Azinphos-ethyl photoproducts, **172**:212
Azinphos-methyl photoproducts, **172**:210
Azinphos-methyl, sublethal AChE inhibition aquatic organisms, **172**:46
Azoxystrobin, hydrolytic profile, **175**:232

Bacteria, capsule-producing, chlorine tolerance, **171**:11
Bacteria, chlorine tolerance by surface attachment, **171**:12
Bacteria, increasing resistance via UV/chlorination, **171**:1 ff.
Bacteria, methyl bromide biodegradation, **177**:61
Bacteria, slime layer-producing, chlorine tolerance, **171**:11
Bacterial aggregation, chlorine exposure, **171**:12
Bacterial chlorine resistance, **171**:2
Bacterial chlorine tolerance, intracellular defenses, **171**:19
Bacterial chlorine tolerance, lab precautions, **171**:13
Bacterial chlorine tolerance, previous chlorination effect, **171**:17
Bacterial endosymbionts, *Acanthamoeba*, **180**:98
Bacterial mechanisms of resistance to antibiotics, **171**:40
Bacterial resistance, antibiotics vs disinfectants, **171**:38
Bacterial resistance, chlorine, **171**:9
Bacterial resistance to antibiotics, UV use, **171**:34
Bacterial resistance to chlorine and antibiotics, **171**:19
Bacterial tolerance, chlorine, **171**:9
Bacterial tolerance to UV light, **171**:30
Baltic Sea (southern), ecosystem changes, **179**:1 ff.
Baltic Sea, a brackish water sea, **179**:2
Baltic Sea drainage area (map), **179**:3
Baltic Sea, heavy metal input sources, **179**:17
Baltic Sea, oxygen content changes, **179**:10
Baltic Sea, PCB sediment levels, **179**:22
Baltic Sea, persistent organic pollutants, **179**:17
Baltic Sea, salinity changes, **179**:10
Baltic Sea, salt content variations, **179**:2
Baltic Sea, sea level increases, **179**:9
Baltic Sea vs Gulf of Gdańsk, comparisons, **179**:6
Baltic Sea water residence time, **179**:2
BART™ (Biological Activity Reaction Tests), **173**:124, 131
Bathochromic effect, photochemistry defined, **172**:134
Bayonet high-pressure mercury arc, described, **172**:145, 146
Baythroid®, cyfluthrin, **174**:73
BCFs (bioconcentration factors), defined, **174**:3
BCFs, pyrethroids, **174**:52, 54
BChE (butyrylcholinesterase), biomarker of OP exposure, **172**:22
BChE activity, parathion-exposed lizards, **172**:43
Beckmann rearrangement, alloxidim sodium, **175**:154
Benomyl, hydrolysis pathway, **175**:118

Benomyl, hydrolytic profile, **175**:200
Bensulfuron methyl, hydrolytic profile, **175**:224
Benzacridines, mutagenicity, **173**:65
Benzenehexachloride (BHC)(HCH), world emission, **173**:93
Benzoin ester formation, pyrethroid hydrolysis, **175**:97
Benzoquinoline metabolism (illus.), **173**:47
Benzoquinolines & metabolites, mutagenicity, **173**:64, 65
Benzoquinolines, genotoxicity, **173**:63
Benzoquinolines, metabolic routes, **173**:44
Benzoylprop-ethyl, hydrolytic profile, **175**:194
Benzoylurea herbicides, structures & hydrolytic profiles, **175**:198
Benzoylurea insecticides, hydrolysis, **175**:107
Benzo[*a*]anthracene (PAH), from coke production, **174**:13
Benzo[*a*]pyrene (BaP), house dust concentrations, **175**:24
Benzo[*a*]pyrene (PAH), from coke production, **174**:13
Benzo[*a*]pyrene, PAH, **179**:75
Benzo[*b*]fluoranthene, PAH, **179**:75
Benzo[*k*]fluoranthene, PAH, **179**:75
Benz[*a*]anthracene, PAH, **179**:75
Beryllium, house dust concentrations, **175**:30
BHC (HCH), world emission, **173**:93
Bifenazate, hydrolytic profile, **175**:232
Bifenthren, physicochemical properties, **174**:52
Bifenthrin, abiotic chemical properties, **174**:68
Bifenthrin, aerobic soil degradation, **174**:73
Bifenthrin, biotic chemical properties, **174**:72
Bifenthrin, hydrolysis in water, **174**:71
Bifenthrin, hydrolytic profile, **175**:192
Bifenthrin, photolysis on soil, **174**:71
Bifenthrin, physicochemical properties, **174**:52, 66
Bifenthrin, soil sorption partition coefficients, **174**:69

Bioaccumulation, cadmium in Antarctic food webs, **171**:64
Bioaccumulation, fipronil, **176**:17
Bioaccumulation, mercury in pelagic seabirds/marine mammals, **171**:69
Bioaccumulation, PFOS (perfluorooctyl sulfonate) in fish, **179**:109
Bioaccumulation, sediment contaminants, **174**:3
Bioaugmentation, pesticide contaminated soils, **177**:170
Bioavailability, metals in soils, defined, **178**:1
Biobeds, disposal pesticide spray spills, **177**:149
Bioconcentration factors (BCFs), defined, **174**:3
Bioconcentration factors (BCFs), fenvalerate, **176**:148
Bioconcentration factors (BCFs), pyrethroids, **174**:52, 54
Bioconcentration factors, PFOS/PFOA, **179**:115
Bioconcentration, PFOS (perfluorooctyl sulfonate) in fish, **179**:111
Biofilms, chromium remediation in wastewater, **178**:142
Biological Activity Reaction Tests (BART™), **173**:124, 131
Biomonitor vertebrates, ranked for utility, **176**:113
Biomonitor vertebrates, ranked for vulnerability to contaminants, **176**:115
Biomonitoring OP exposure, aquatic environments, **172**:42
Bioremediation technology, Cr(VI) in soils, **178**:144
Biostimulation, pesticide contaminated soils, **177**:178
Biosuper effect, sulfur microbial oxidation effect soil, **177**:12
Biotic coupling, manganese oxidation of Cr(III), **178**:121
Biotransformation, PFAS, **179**:111
Bird toxicity, avermectins, **171**:126
Birds, arsenic effects, **180**:146
Birds, behavioral disturbances OP exposure, **172**:37
Birds, blood ChE & CbE biomarkers OP exposure, **172**:37

Birds, ChE recovery times OP exposure, **172**:26
Birds, fipronil toxicity, **176**:32
Birds, proposed arsenic criteria, **180**:157
Birds, range of OP toxicity studies, **172**:36
Bismuth, house dust concentrations, **175**:30
Black plug layering consortium (microbial), **173**:137
Black River, brown bullheads liver lesions, **174**:13
Black River, PAH-sediment remediation, **174**:12
Blood cholinesterases (ChE), biomarker OP exposure, **172**:22
Boiling points, indoor pollutants (table), **175**:2
Boltzmann's constant, abiotic pesticide hydrolysis, **175**:83
Bone formation, teleosts, **172**:10
Bone tissue, composition, **172**:10
Boranes, polychlorinated, chiral insecticide, **173**:91
Boron, house dust concentrations, **175**:29
Brackish water, Baltic Sea, **179**:2
Brain AChE vs serum ChE activity, lizards OP-exposed, **172**:51
Bromine residues, above-ground plant tissues, methyl bromide, **177**:66
Bromine residues, edible plant parts, methyl bromide, **177**:67
Bromine residues, plant seedlings, methyl bromide, **177**:66
Bromine residues, soils/plants from methyl bromide fumigation, **177**:62, 64
Bromine vs chlorine, ozone breakdown efficiency, **177**:47
Bromocyclen, chiral insecticide, **173**:91
Bromomethane, see methyl bromide, **177**:46
Bromophos, hydrolytic profile, **175**:208
Buffer, abiotic pesticide hydrolysis, **175**:84
Buffer catalysis, pesticide hydrolysis in aquatic systems, **175**:164
Bumblebee adults, spinosad toxicity, **179**:43

Bumblebee larvae, spinosad toxicity, **179**:51
Butamifos, hydrolytic profile, **175**:220
Butox®, deltamethrin, **174**:95
Butyrylcholinesterase (BChE), biomarker of OP exposure, **172**:22

Cadmium, algal toxicity (table) , **178**:38
Cadmium, bioaccumulation field crops, **177**:2
Cadmium, bioaccumulation in Antarctic food webs, **171**:64
Cadmium, concentration in fertilized/unfertilized soils Australia, **177**:19
Cadmium concentrations, Antarctic fish, **171**:53 ff., 63
Cadmium concentrations, seabirds Southern Ocean, **171**:67
Cadmium fixation in soils, **178**:5
Cadmium, house dust concentrations, **175**:29
Cadmium, occupational hair levels, **175**:58
Cadmium, soil accumulation from fertilizers, **177**:14
Cadmium, soil contaminant sources, **177**:3
Cadmium transfer to birds/seals breeding in Antarctica, **171**:77
Calcitonin, teleost-secreted, **172**:5
Calcium dihydrogen phosphate, fertilizer, **177**:11
Calcium monohydrogen phosphate, fertilizer, **177**:11
Calcium/phosphate regulation, pesticide effects teleosts, **172**:8
Cancer, various forms arsenic-related, **180**:136
Capsule-producing bacteria, chlorine tolerance, **171**:11
Captafol, hydrolytic profile, **175**:204
Captan, hydrolytic profile, **175**:204
Carassius auratus, chlorinated tissue residues, **173**:12
Carbamate pesticides, hydrolysis kinetics mechanisms, **175**:108
Carbamate pesticides, structures & hydrolytic profiles, **175**:200
Carbamate potentiation, by OPs, **172**:39
Carbamates, acid and alkaline hydrolysis mechanisms, **175**:110

Carbaryl, hydrolytic profile, **175**:200
Carbazole & homocyclic analog fluorene (illus.), **173**:41
Carbofuran, degradation by photo-Fenton process, **177**:160
Carbofuran, hydrolytic profile, **175**:200
Carbonate apatite, phosphate fertilizer, **177**:11
Carboxin, hydrolytic profile, **175**:196
Carboxylesterase (CbE), role in OP toxicity, **172**:31
Carboxylic esters, hydrolysis, **175**:90
Carcinogenicity, azaarenes, **173**:62
Carcinogenicity, chromium, **178**:59
Carcinogenicity, DINP, **172**:108
Carcinogenicity, fipronil, **176**:35, 45
Carfentrazone hydrolysis, **175**:162
Carfentrazone, hydrolytic profile, **175**:232
Cartap, hydrolysis pathway, **175**:119
Cartap, hydrolytic profile, **175**:200
CAS chemical registration numbers, see Chemical Abstract regis nos, **174**:66
CAS numbers, chiral pesticides, **173**:107
CAS numbers, drugs (table), **180**:76
CAS numbers, organophosphate insecticides, **172**:215
CAS numbers, pesticide list, **177**:183
CAS numbers, pesticides veterinary use (table), **180**:76
CAS numbers, veterinary medicines (table), **180**:76
Cation exchange capacities (CECs), soils, **178**:4
CbE (carboxylesterase), role in OP toxicity, **172**:31
CbE activity, parathion-exposed lizards, **172**:43
CBR approach, uncertainties, **173**:6
CBR defined, **173**:2
CBR determinants, **173**:10
CBR evaluation, methodology, **173**:6
CBR variability, **173**:9
CBRs (Critical Body Residues), **173**:2
CBRs, acute/chronic toxicity of chemical classes, aquat organ, **173**:5
CBRs, ranges affecting survival, **173**:27
CCU (urea herbicide), hydrolytic profile, **175**:198
CECs (See cation exchange capacities, **178**:4

Cellular oxygen uptake, chlorine effects, **171**:7
Cellular uptake, chlorine effects, **171**:6
Cephalosporins, mechanisms of action, **171**:39
Chamber methods, methyl bromide volatilization, **177**:83
Charcoal tubes, methyl bromide adsorbent, **177**:57
ChE (blood cholinesterases), biomarker of OP exposure, **172**:22
ChE, biomarker of OP exposure, aquatic invertebrates, **172**:23
ChE I$_{50}$s, OP concentrations various wildlife (table), **172**:28
ChE inhibition, sublethal OP levels aquatic organisms, **172**:44
ChE measurements, aquatic invertebrates (table), **172**:24
ChE recovery times, OP exposure aquatic invertebrates, **172**:25
ChE recovery times, OP exposure birds, **172**:26
ChE recovery times, OP exposure fish, **172**:25
ChE recovery times, OP exposure non-mammalian vertebrates, **172**:25
ChE recovery times, OP exposure reptiles, **172**:26
Chelators, effect in algal toxicity testing, **178**:30
Chemical Abstracts registration numbers, pyrethroids, **174**:66, 73, 82, 95, 117, 125, 139, 151
Chemical Abstracts Service Registry numbers, OP insecticides, **172**:215
Chemical classes, modes of action, **173**:8
Chemical names, drugs (table), **180**:76
Chemical names, organophosphate insecticides, **172**:215
Chemical names, pyrethroids, **174**:66, 73, 82, 95, 117, 125, 139, 151
Chemical names, veterinary medicines (table), **180**:76
Chemical properties, pyrethroids, **174**:49 ff.
Chemical reactivity, chromate ion, **178**:57
Chemical structure, triclopyr, **174**:20
Chemical structures, fenvalerate & esfenvalerate, **176**:139

Chemical warfare agents, National Research Council (NRC) Review, **172:** 71
Chemical warfare agents, oral reference doses, **172:**65 ff.
Chemical warfare agents, RfDs & total uncertainty factors (UFs), **172:**68
Chemical warfare agents, toxicological criteria for RfDs, **172:**80
Childhood exposure, diisononyl phthalate, **172:**111
Chiral compounds, enrichment processes, **173:**105
Chiral compounds, ERs in different compartments, **173:**93
Chiral compounds, mirror image structures, **173:**92
Chiral compounds, shielding from the racemate, **173:**104
Chiral PCBs, **173:**91
Chiral pesticides, constant enantiomer fraction, **173:**101
Chiral pesticides, enantiomeric enrichment in environ, **173:**85 ff.
Chiral pesticides, enrichment processes, **173:**105
Chiral pesticides, list, **173:**90
Chiral pesticides, racemic mixture deviations, **173:**100
Chiral pesticides, shielding from the racemate, **173:**104
Chiral pesticides, worldwide use, **173:**86
Chlamydomonas reinhardii (alga), pH effects Cu toxicity, **178:**30
Chloramphenicol, mechanisms of action, **171:**39
Chloramphenicol resistant bacteria, **171:** 21, 23
cis-Chlordane, CAS number, **173:**107
cis-Chlordane, chiral insecticide, **173:**90, 97, 104
cis-Chlordane, enantiomer structures, **173:** 92
cis-Chlordane, IUPAC chemical name, **173:**107
trans-Chlordane, CAS number, **173:**107
trans-Chlordane, chiral insecticide, **173:** 90, 98, 104
trans-Chlordane, enantiomer structures, **173:**92

trans-Chlordane, IUPAC chemical name, **173:**107
Chlordane compounds, chiral insecticides, **173:**90, 94
Chlordane compounds, stereochemical recognition, **173:**103
Chlordane, house dust concentrations, **175:**19
Chlordane-*cis*, chiral insecticide, **173:**90, 97, 104
Chlordane-*trans*, chiral insecticide, **173:** 90, 98, 104
Chlordimeform, hydrolytic profile, **175:** 232
Chlorella pyrenoidosa (alga), Cu & Pb testing, **178:**28
Chlorella vulgaris (alga), toxicity testing, **178:**34, 37
Chlorfenvinphos, hydrolytic profile, **175:** 212
Chlorfenvinphos photoproducts, **172:**153
Chlorimuron ethyl, hydrolytic profile, **175:**224
Chlorinated pesticides, fish, **179:**27
Chlorination, increasing resistance in bacteria, **171:**1 ff.
Chlorine/antibiotic resistance in bacteria, **171:**19
Chlorine, bacterial resistance, **171:**2
Chlorine, bacterial tolerance or resistance?, **171:**9
Chlorine disinfection today, **171:**25
Chlorine, dosages used for various applications (table), **171:**3
Chlorine, effect on amino acids, **171:**6
Chlorine, effect on ATP formation, **171:**8
Chlorine, effect on cellular uptake, **171:**6
Chlorine, effect on DNA, **171:**8
Chlorine, effect on membrane permeability, **171:**6
Chlorine, effect on sulfhydryl enzymes, **171:**6
Chlorine, history in water disinfection, **171:**2
Chlorine, mechanisms of action, **171:**4, 5
Chlorine, mechanisms of tolerance bacteria, **171:**10
Chlorine, promotes bacterial aggregation, **171:**12
Chlorine resistance, bacteria, **171:**19

Chlorine tolerance, bacteria attach to surfaces, **171**:13
Chlorine tolerance, bacterial antecedent growth, **171**:14
Chlorine tolerance, encapsulated bacteria, **171**:11
Chlorine tolerance, previous exposure effects, **171**:17
Chlorine, use as disinfectant, **171**:4
Chlorine, used as disinfectant by nature, **171**:3
Chlorine-resistant bacteria, mechanisms of tolerance, **171**:10
Chlormephos photoproducts, **172**:196
Chloropicrin, added warning agent in methyl bromide, **177**:49
Chloropicrin, methyl bromide replacement, **177**:107
Chloroxuron, hydrolytic profile, **175**:198
Chlorpropham, hydrolytic profile, **175**:200
Chlorpyrifos, house dust concentrations, **175**:19, 21
Chlorpyrifos, hydrolytic profile, **175**:212
Chlorpyrifos, metabolic pathway (diag.), **172**:3
Chlorpyrifos, metabolite chemical structures, **172**:3
Chlorpyrifos methyl, hydrolytic profile, **175**:212
Chlorpyrifos oxon, **172**:3
Chlorpyrifos photoproducts, **172**:185
Chlorpyrifos, sublethal AChE inhibition aquatic organisms, **172**:44
Chlorpyrifos-oxon I_{50}s, ChE fish, **172**:29
Chlorsulfuron, hydrolytic profile, **175**:224
Chlorthion, hydrolytic profile, **175**:208
Chlozolinate, hydrolytic profile, **175**:204
Cholinesterase-inhibiting pesticides, utility index & score, **176**:82, 86
Cholinesterase-inhibiting pesticides, vertebrate vulnerability index, **176**:81, 83, 87
Chromate ion, chemical reactivity, **178**:57
Chromate-contaminated soil/water, remediation, **178**:77
Chromate-contaminated soils, remediation technologies, **178**:79
Chromite, only major commercial product, **178**:95
Chromium (III), chemistry, **178**:57
Chromium (III), oxidation rate by γ-MnOOH, **178**:102
Chromium (III), threshold permissible levels in soils, **178**:78
Chromium (V), accumulation in Cr(VI) reduction to Cr(III), **178**:119
Chromium (VI), bioremediation technology, **178**:144
Chromium (VI), chemistry, **178**:57
Chromium (VI) direct reduction, microbial, **178**:104
Chromium (VI), factors affecting soil solution, **178**:72
Chromium (VI) reductases, bacteria, **178**:118
Chromium (VI) reduction, list of capable microbes, **178**:106
Chromium (VI) reduction microbes, list, **178**:106
Chromium (VI) reduction, microbial aerobically, **178**:106
Chromium (VI) reduction, microbial anaerobically, **178**:106
Chromium (VI), soil adsorption, **178**:74
Chromium (VI), threshold permissible levels in soils, **178**:78
Chromium (VI) tolerance, microbes, **178**:103
Chromium (VI) transformation to nontoxic Cr(III), **178**:94
Chromium (VI)-tolerant bacteria, **178**:125
Chromium, adsorption/desorption, soils, **178**:70
Chromium, anthropogenic sources, **178**:95
Chromium, carcinogenicity, **178**:59
Chromium chemistry, soils, **178**:53 ff.
Chromium, commercial uses, **178**:54
Chromium cycle in soil and water (diagram), **178**:60
Chromium, effects on enzymes in soils, **178**:138
Chromium, effects on microbial processes in soils (table), **178**:133
Chromium, effects on nitrogen fixation in soil, **178**:139
Chromium, effects on pure cultures, microorgs (table), **178**:133

Chromium, effects on reductive dechlorination in soil, **178**:140
Chromium, effects on soil dehydrogenase activity, **178**:132
Chromium, effects on soil microbial communities, **178**:130
Chromium, effects on soil respiration, **178**:139
Chromium extraction, serpentine soils, **178**:62
Chromium extraction, sludge, **178**:62
Chromium genotoxicity, **178**:59
Chromium, house dust concentrations, **175**:29
Chromium, in soil solution, **178**:67
Chromium, in tannery waste sites, **178**:53 ff.
Chromium, industrial consumption, **178**:96
Chromium, inhibitory effects on soil nitrification, **178**:139
Chromium, iron effects on oxidation, **178**:98
Chromium, irreversible reduction, Cr(VI) to Cr(III), **178**:78
Chromium, manganese effects on oxidation, **178**:99
Chromium, natural sources, **178**:95
Chromium, occupational hair levels, **175**:58
Chromium, organically complexed solubility, **178**:69
Chromium, oxidation of Cr(III) to Cr(VI), **178**:71
Chromium oxidation reactions, soil, **178**:98
Chromium oxidation, soil type effects, **178**:100
Chromium, partitioning & mobility, **178**:66
Chromium partitioning, sludge/soils, **178**:63
Chromium, priority pollutant, **178**:94
Chromium, reduction potential diagram, **178**:56
Chromium remediation in soils, **178**:143
Chromium remediation in wastewater, **178**:141
Chromium salts, **178**:54

Chromium sequential extraction scheme, Chinese soils, **178**:63
Chromium, soil contaminant sources, **177**:3
Chromium, soil mobility, water solubility, **178**:97
Chromium, soil organic matter effects, **178**:97
Chromium, soil solution chemistry, **178**:67
Chromium, solid-phase speciation, **178**:61
Chromium, stable oxidation states, **178**:55
Chromium, thermodyn stability, aqueous species (diag), **178**:68
Chromium toxicity, algae, **178**:129
Chromium toxicity, animals, humans, **178**:58
Chromium toxicity, fungi, **178**:127
Chromium toxicity, general, **178**:58
Chromium toxicity, microorganisms, **178**:126, 128
Chromium toxicity, mycorrhizal fungi, **178**:127
Chromium toxicity, plants, microorganisms, **178**:58
Chromium transformation in soil, factors governing, **178**:97
Chromium transformation in soil, microbe factors, **178**:103
Chromium, various forms, **178**:95
Chromium, water solubility, **178**:97
Chromium-contaminated soils, microorganism remediation, **178**:93 ff.
Chromium-contaminated soils, sources, **178**:94
Chromium-contaminated water & soils, remediation, **178**:141
Chromium-microorganism interactions, soil, **178**:93 ff.
Chromophores, spectral absorption maxima (table), **172**:136
Chromosome aberrations, DINP, **172**:108
Chronic toxicity, fipronil mammals, **176**:34
Chrysene, PAH, **179**:75
CIRAD-GERDAT-PRIFAS, fipronil tests, **176**:40
Circumpolar biomonitoring networks, **171**:53 ff.

Cislin®, deltamethrin, **174**:95
Clams, trace metals Antarctica, **171**:83
Class 1 and 2 pyrethroids, hydrolysis, **175**:95
Cleanup of pesticide contaminated soils, **177**:161
CNS seizure agents, CBRs, **173**:5
CNS seizure agents, defined, **173**:2
CNS seizure agents, mode of action, **173**:8
CNS seizure agents, tissue residues, **173**:19
Coastal benthic invertebrates, trace metal levels (Ross Sea), **171**:75
Coastal benthic organisms, trace metal levels, **171**:73
Cobalt fixation in soils, **178**:6
Cobalt, house dust concentrations, **175**:30
Cobalt, occupational hair levels, **175**:58
Coke production, PAH source, **174**:12
Collagen content, fish response to pesticides, **172**:12
Commission on Agrochemicals and the Environment, **177**:125
Common names, terrestrial vertebrates, **176**:70
Common names, organophosphate insecticides, **172**:215
Comparative toxicity, azaarenes vs homocyclic PAHs, **173**:67
Complexation affinity, metal cations by organic matter (diagram), **177**:7
Composting, pesticide contaminated soils, **177**:166
Consortial microbial events, deep sea rusticles, **173**:117 ff.
Consortial nature of microbial events, *Titanic*, **173**:117 ff.
Contact lens wearers, risk factors keratitis, **180**:122
Contact lenses, *Acanthamoeba* keratitis association, **180**:119
Contact lenses, age distribution of users, **180**:120
Contact lenses, demographics of use, **180**:119
Contact lenses, eye disease association, **180**:119

Contact lenses, history of development, **180**:119
Contact lenses, types, **180**:120
Container and packaging (pesticide) disposal, **177**:142
Contaminant effects, aquatic organisms, **173**:1 ff.
Contaminated sediment, aquatic environment effects, **174**:2
Contaminated sediment remediation, **174**:1 ff.
Contaminated sediment removal, dredging, **174**:10
Contaminated sediments, aquatic ecosystem effects, **174**:4
Cooking ware, hair trace element contamination, **175**:55
Copper concentrations, Antarctic fish, **171**:53 ff., 63
Copper fixation in soils, **178**:5
Copper, occupational hair levels, **175**:58
Copper, soil contaminant sources, **177**:3
Copper toxicity, earthworms, **178**:10
Cosmetics, hair trace element contamination, **175**:59
Cotton glove, house dust press sampling, **175**:10
Cottonseed, aflatoxins Arizona monitoring results, **171**:146
Cottonseed, ammonia treatment of aflatoxin-contaminated, **171**:148
Coumaphos, hydrolytic profile, **175**:212
Coumaphos photoproducts, **172**:192
Crayfish, bioindicators of water pollution, **172**:23
Critical Body Residues (CBRs), **173**:2
Croococcus paris (alga), zinc toxicity testing, **178**:37
Cross-resistance (insect), fipronil, **176**:3, 52
Culture media, algal toxicity testing, **178**:28
Cyanazine, hydrolytic profile, **175**:230
Cyanofenphos, hydrolytic profile, **175**:220
Cyanogen chloride, RfDs & UFs, **172**:68
Cyanophos photoproducts, **172**:180
Cyclic dicarboximide fungicides, hydrolysis pathways, **175**:122

Cyclic dicarboximide fungicides, structures & hydrolytic profiles, **175**:204
Cyclic dicarboximide pesticides, hydrolysis mechanisms, **175**:120
Cyclotella sp. (alga), toxicity testing, **178**:34
Cyfluthrin, abiotic chemical properties, **174**:77
Cyfluthrin, aerobic soil degradation, **174**:80
Cyfluthrin, Baythroid®, **174**:73
Cyfluthrin, biotic chemical properties, **174**:80
Cyfluthrin, hydrolysis, **174**:77, 79
Cyfluthrin, hydrolytic profile, **175**:192
Cyfluthrin, photolysis in water, **174**:77, 80
Cyfluthrin, physicochemical properties, **174**:52, 73
Cyfluthrin, soil sorption partition coefficients, **174**:78
Cyhalothrin, degradation in soil (chart), **174**:59, 60
Cyhalothrin, hydrolytic profile, **175**:192
Cypermethrin, abiotic chemical properties, **174**:87
Cypermethrin, aerobic soil degradation, **174**:92
Cypermethrin, aquatic degradation, **174**:92, 94
Cypermethrin, hydrolysis, **174**:87, 90
Cypermethrin, hydrolytic profile, **175**:192
Cypermethrin, photolysis in water, **174**:87, 91
Cypermethrin, photolysis on soil, **174**:90
Cypermethrin, physicochemical properties, **174**:52, 82
Cypermethrin, soil sorption partition coefficients, **174**:86, 88

2,4-D, house dust contaminant, **175**:16
2,4-D, hydrolytic profile, **175**:190
2,4-D, soil leaching qualities, **174**:30
DANIDA report, fipronil, **176**:42
Daphnia magna, antibacterials toxicity, **180**:71
Daphnia magna, narcotic tissue residues, **173**:11
2,4'-DDT, chiral insecticide, **173**:91

2,4'-DDD, chiral insecticide, **173**:91
p,p'-DDT, hydrolytic profile, **175**:206
DDT, herring, Baltic Sea, **179**:28
DDT, house dust concentrations, **175**:19
DDT residues, mummichogs, **173**:21
DDT-contaminated sediment, Great Lakes, **174**:12
Decis®, deltamethrin, **174**:95
Deet, hydrolytic profile, **175**:196
Degradation pathway (diagram), triclopyr, **174**:25
Degradation pathways, fenvalerate soils, **176**:145
Degradation pathways, fipronil, **176**:12, 13
Dehalogenation, pesticide waste disposal, **177**:151
Dehydrogenases, chromium effects in soil, **178**:132
Deinococcus spp., radiation-resistant bacteria, **173**:134
Deltamethrin, abiotic chemical properties, **174**:104
Deltamethrin, aerobic aquatic degradation, **174**:113, 116
Deltamethrin, aerobic soil degradation, **174**:112
Deltamethrin, biotic chemical properties, **174**:111
Deltamethrin, Butox®, Cislin®, Decis®, K-Othrin®, **174**:95
Deltamethrin, hydrolysis, **174**:104, 108
Deltamethrin, hydrolytic profile, **175**:192
Deltamethrin, photolysis in water, **174**:106, 109
Deltamethrin, photolysis on soil, **174**:110
Deltamethrin, physicochemical properties, **174**:52, 95
Deltamethrin, soil sorption partition coefficients, **174**:100, 103
Demeton-S, hydrolytic profile, **175**:212
Demeton-S sulphone, hydrolytic profile, **175**:212
Demeton-S sulphoxide, hydrolytic profile, **175**:212
Demeton-S-methyl, hydrolytic profile, **175**:212
Demeton-S-methyl sulphone, hydrolytic profile, **175**:212

Denitration, new photochemical behavior, **172**:173
Dermal toxicity, fipronil, **176**:35
Des-ethyl chlorpyrifos, metabolite, **172**:3
Developmental toxicity, DINP, **172**:106
Dialkyl phthalate (DAPs), plastic softener uses, **172**:87
Diazinon, hydrolytic profile, **175**:208
Diazinon I_{50}s, ChE fish, **172**:29
Diazinon photoproducts, **172**:189
Diazoxon I_{50}s, ChE fish, **172**:29
Dibenzacridines, mutagenicity, **173**:66
Dibenz[a,h]anthracene, PAH, **179**:75
Dicamba, soil leaching index, **174**:33
Dicapthon, hydrolytic profile, **175**:208
Dicarboximide fungicides, hydrolysis pathways, **175**:122
Dicarboximide fungicides, structures & hydrolytic profiles, **175**:204
Dicarboximide pesticides, hydrolysis mechanisms, **175**:120
Dichloropropene (1,3-D), methyl bromide replacement, **177**:107
Dichlorprop, chiral herbicide, **173**:90
Dichlorvos, hydrolytic profile, **175**:212
Dichlorvos I_{50}s, ChE fish, **172**:28
Dichlorvos I_{50}s, ChE fish, **172**:28
Diclofop-methyl, hydrolytic profile, **175**:190
Dicofol, hydrolytic profile, **175**:206
Dicrotophos, hydrolytic profile, **175**:212
Diet effect, hair trace element contamination, **175**:54
Diethanyl-ethyl, hydrolytic profile, **175**:194
Diflubenzuron, hydrolysis pathways, **175**:109
Diflubenzuron, hydrolytic profile, **175**:198
Diflufenican, hydrolytic profile, **175**:232
Diisononyl phthalate (DINP), chemistry, **172**:87 ff.
Diisononyl phthalate (DINP), environmental fate, **172**:87 ff.
Diisononyl phthalate (DINP), toxicology, **172**:87 ff.
Diisononyl phthalate, analytical methods, **172**:93
Diisononyl phthalate, end uses (table), **172**:91
Diisononyl phthalate, environmental concerns, **172**:93
Diisononyl phthalate, environmental levels, **172**:95
Diisononyl phthalate, health concerns, **172**:88
Diisononyl phthalate, mode of action, **172**:109
Diisononyl phthalate, physicochemical properties, **172**:89
Diisononyl phthalate, production & use, **172**:89
Diisononyl phthalate, sources, **172**:88
Dimer formation, allethrin alkaline hydrolysis, **175**:94
Dimethoate, sublethal AChE inhibition aquatic organisms, **172**:45
Dimetilan, hydrolytic profile, **175**:200
Dinitroaniline herbicides, chemical structures, **175**:168
DINP (diisononyl phthalate), environmental fate, **172**:87 ff.
DINP, acute toxicity mammals, **172**:101
DINP, analytical methods, **172**:93
DINP, biological effects, **172**:97
DINP, carcinogenicity, **172**:108
DINP, childhood exposure, **172**:111
DINP, developmental toxicity, **172**:106
DINP, end uses (table), **172**:91
DINP, endocrine modulation, **172**:110
DINP, environmental concerns, **172**:93
DINP, environmental levels, **172**:95
DINP, environmental sampling techniques, **172**:94
DINP, environmental transformations/fate, **172**:96
DINP, exposure assessment, **172**:115
DINP, flexible PVC plasticizer, **172**:91
DINP, hazard characterization, **172**:113
DINP, human exposure, **172**:111
DINP, human health risk assessment, **172**:111
DINP, mammalian toxicity, **172**:99
DINP, metabolism mammals, **172**:99
DINP, mode of action, **172**:109
DINP, mutagenicity/genotoxicity, **172**:107
DINP, occupational exposure, **172**:111
DINP, release to environment, **172**:93
DINP, reproductive effects, **172**:103

DINP, tolerable daily intake, **172**:114
DINP, toxicity aquatic organisms, **172**: 98
DINP, toxicology, **172**:97
DINP, toy manufacturing, **172**:92
DINP, uses in plastics industry, **172**:87
Dioxabenzophos photoproducts, **172**:195
Dioxins (PCDDs), house dust contaminants, **175**:27
Dislodgeable residues, spinosad toxicity, **179**:44
Disposal of unused pesticide stocks, **177**:134
Dissolved organic carbon, effects on metal aging, soils, **178**:5
Dissolved organic matter, abiotic pesticide hydrolysis, **175**:86
Dithiophosphates, organophosphates photochemistry, **172**:195
Diuron, hydrolytic profile, **175**:198
DMPA (Zytron®), hydrolytic profile, **175**:220
DNA, chlorine effects, **171**:8
DOC (See dissolved organic carbon), **178**:5
Domestic animals, arsenic effects, **180**:148
Domestic livestock, proposed arsenic criteria, **180**:157
DPSIR, identified, **179**:30
DPSIR, illustrated, **179**:33
Dredging, removal of contaminated sediments, **174**:10
Drinking water, proposed arsenic criteria, **180**:156
Drinking water residues, fipronil, **176**:19
Drug aquatic toxicity, **180**:44 ff.
Drug degradation, veterinary different scenarios (data), **180**:37
Drug disposal, veterinary unwanted, **180**:16
Drug terrestrial toxicity, **180**:63 ff.
Drugs, CAS numbers (table), **180**:76
Drugs, chemical names (table), **180**:76
DTPA (See diethylenetriaminepentaacetic acid), **178**:7
DTPA, metal extraction, soils, **178**:7
Dung fauna, ivermectin effects, **171**:117
Dye-sensitized photooxidation, organophosphate photochemistry, **172**:199

Earthworms, avermectin effects, **171**:121, 123
EBI (ergosterol biosynthesis inhibiting) fungicides, **172**:39
Economic impact, methyl bromide phase-out, **177**:47
Ecosystem changes, Gulf of Gdańsk, **179**:1 ff.
EDTA (See ethylenediaminetetraacetic acid), **178**:7
EDTA, metal extraction, soils, **178**:7
Eels, euryhaline, **172**:6
EF (enantiomer fractions), defined, **173**:88
Egyptian phosphate rock, fertilizer, **177**:20
Electrochemical fluorination, reaction described, **179**:100
Electromagnetic radiation spectrum (diag.), **172**:130
Electron transport chain, chlorine effects, **171**:7
ELISAs (enzyme-linked immunosorbent assays), house dust, **175**:15
Emamectin benzoate, environmental metabolites, **180**:17
Emission comparisons, photochemical lamps, **172**:143
Emission spectrometry, house dust contaminant analysis, **175**:14
Enantiomer fractions (EF), defined, **173**:88
Enantiomer fractions (EFs), chiral compounds, **173**:102
Enantiomeric enrichment, chiral pesticides in environment, **173**:85 ff.
Enantiomeric enrichment, methodology, **173**:88
Enantiomeric ratio (ER), defined, **173**:86, 88
Encapsulated bacteria, chlorine tolerance, **171**:11
Endocrine disruption, pyrethroids, **176**:152
Endocrine modulation, DINP, **172**:110
Endocrine-disrupting agents, house dust contaminants, **175**:26
Endocrine-disrupting effects, veterinary medicines, **180**:73
Endosulfan, hydrolytic profile, **175**:206

Ene-type reaction, UV affects nitroaryl compounds, **172:**162
Environmental bioavailability, metals in soils, **178:**9
Environmental contaminants, vertebrates ranked as biomonitors, **176:**67 ff.
Environmental contamination, veterinary medicines data, **180:**18 ff.
Environmental effects, hair trace element contamination, **175:**60
Environmental effects, perfluoroalkylated substances (PFAS), **179:**99 ff.
Environmental fate, methyl bromide fumigant, **177:**45 ff.
Environmental fate processes, methyl bromide, **177:**58
Environmental metals risk testing, algae, **178:**23 ff.
Environmental monitoring, veterinary medicines, **180:**18 ff.
Environmental no effect concentration, PAHs, **179:**83
Environmental pollutants, vertebrate vulnerability indices, **176:**72
Environmental risk assessment, avermectins, **171:**127
Enzyme-linked immunosorbent assays (ELISAs), house dust analysis, **175:**15
EPA pesticide honeybee toxicity groups, **179:**66
Epicuticular waxes, effects on OP photodegradation, **172:**182
EPN, hydrolytic profile, **175:**220
ER (enantiomeric ratio), defined, **173:**86, 88
ER, as a tracer tool in environmental studies, **173:**87
Ergosterol biosynthesis inhibiting (EBI) fungicides, procloraz & penconazole, **172:**39
Esfenvalerate, abiotic chemical properties, **174:**121
Esfenvalerate, aerobic aquatic degradation, **174:**127
Esfenvalerate, aerobic soil degradation, **174:**123, 126
Esfenvalerate, Asana®, **174:**117
Esfenvalerate, biotic chemical properties, **174:**123

Esfenvalerate, chemistry & fate, **176:**137 ff.
Esfenvalerate, hydrolysis, **174:**121
Esfenvalerate hydrolytic profile, **175:**192
Esfenvalerate, photolysis in water, **174:**121
Esfenvalerate, photolysis on soil, **174:**122
Esfenvalerate, physicochemical properties, **174:**52, 117
Esfenvalerate, physicochemical properties, **176:**138, 140
Esfenvalerate, uses, **176:**141
Essential elements, plant/animal nutrition, defined, **177:**2
Ethametsulfuron methyl, hydrolytic profile, **175:**224
Ethiofencarb, hydrolytic profile, **175:**200
Ethion, hydrolytic profile, **175:**220
Ethyl parathion, hydrolytic profile, **175:**208
Ethylenebiscarbamate, hydrolytic profile, **175:**200
ETox model, veterinary medicines, **180:**7
Euphausia superba, trace metal concentrations Southern Ocean, **171:**63
EUROCAT project, marine pollution control, **179:**30
Eutrophication, Baltic's major ecological problem, **179:**2
Eutrophication, Gulf of Gdańsk, **179:**1 ff.
Excitatory agents, CBRs, **173:**5
Excitatory agents, mode of action, **173:**8
Excitatory agents, tissue residues, **173:**15
Excited state, photon absorption, defined, **172:**131
Exposure assessment, diisononyl phthalate, **172:**115
Exposure assessment models, veterinary medicines, **180:**8
EXTOXNET website, **172:**48
Extraction methods, chromium in soil, **178:**62
Extrapolation, laboratory to field, OP exposure, **172:**39
Eye infections, *Acanthamoeba* keratitis, **180:**104

Famoxadone, hydrolytic profile, **175:**204
FDA action levels, aflatoxins in animal feeds, **171:**143

FDA action levels, aflatoxins in milk, **171**:144
FDA, aflatoxins feedstuffs surveillance results, **171**:146
Feces contamination, veterinary medicines, **180**:22
Feeding studies, aflatoxin-ammoniation of contaminated feeds, **171**:160
Fenitrooxon I_{50}s, ChE fish, **172**:28
Fenitrothion, hydrolytic profile, **175**:208
Fenitrothion photoproducts, **172**:170
Fenoxaprop-ethyl, hydrolytic pathway, **175**:93
Fenoxaprop-ethyl, hydrolytic profile, **175**:190
Fenpropathrin, abiotic chemical properties, **174**:134
Fenpropathrin, aerobic aquatic degradation, **174**:138
Fenpropathrin, aerobic soil degradation, **174**:136
Fenpropathrin, anaerobic aquatic degradation, **174**:139
Fenpropathrin, hydrolysis, **174**:134
Fenpropathrin, hydrolytic profile, **175**:192
Fenpropathrin, photolysis in water, **174**:134
Fenpropathrin, photolysis on soil, **174**:136
Fenpropathrin, physicochemical properties, **174**:52, 125
Fenpropathrin, soil sorption, **174**:129, 132
Fensulfothion, hydrolytic profile, **175**:208
Fenthion, hydrolytic profile, **175**:208
Fenthion photoproducts, **172**:181
Fenuron, hydrolytic profile, **175**:198
Fenvalerate, bioconcentration factors (BCFs), **176**:148
Fenvalerate, biotransformation pathway, **176**:150
Fenvalerate, chemical structure(s), **176**:139
Fenvalerate, chemistry & fate, **176**:137 ff.
Fenvalerate, degradation pathway in soils, **176**:145
Fenvalerate, fate in aqueous environments, **176**:144
Fenvalerate, fate in soils, **176**:144
Fenvalerate, hydrolytic profile, **175**:192
Fenvalerate, metabolic pathway in animals, **176**:150

Fenvalerate, photodegradation pathways, **176**:147, 149
Fenvalerate, photolysis to decarboxyfenvalerate, **176**:146
Fenvalerate, physicochemical properties, **176**:138, 140
Fenvalerate, Pydrin®, **174**:117
Fenvalerate residues, rainbow trout, **173**:21
Fenvalerate, sorption to sediments, **176**:145
Fenvalerate, synthesis, **176**:141, 142
Fenvalerate, uses, **176**:141
FETAX (Frog embryo teratogenesis assay–*Xenopus*), **172**:35
Field experiments, methyl bromide volatilization, **177**:89
Fiprole insecticides, defined, **176**:2
Fipronil, acute toxicity, **176**:34
Fipronil, ADI, **176**:46
Fipronil, adverse effects in domestic animals, **176**:44
Fipronil, aerobic soil metabolism, **176**:15
Fipronil amide, **176**:14
Fipronil, anaerobic aquatic metabolism, **176**:15
Fipronil, bioaccumulation, **176**:17
Fipronil, carcinogenicity, **176**:35, 45
Fipronil, chemical structure, **176**:6
Fipronil, chronic toxicity, **176**:34
Fipronil, degradation pathway soil, **176**:13
Fipronil, degradation rates soil, **176**:15
Fipronil, degradation routes rice fields, **176**:12
Fipronil, dermal toxicity, **176**:35
Fipronil desulfinyl, **176**:14
Fipronil desulfinyl, toxic photodegradate, **176**:48
Fipronil detrifluoromethylsulphinyl, **176**:14
Fipronil, developmental toxicity, **176**:36
Fipronil, ecotoxicology, **176**:1 ff.
Fipronil, effects aquatic invertebrates, **176**:24, 25
Fipronil, effects aquatic organisms, **176**:23, 25
Fipronil, effects birds, **176**:31
Fipronil, effects fish, **176**:25
Fipronil, effects lizards, **176**:30

Fipronil, effects microorganisms, **176**:26
Fipronil, effects plants, **176**:26
Fipronil, effects terrestrial invertebrates, **176**:27
Fipronil, environmental fate, **176**:1 ff., 10
Fipronil, formulations, **176**:7
Fipronil, genotoxicity, **176**:36
Fipronil, human adverse reactions, **176**:44
Fipronil, human health concerns, **176**:1 ff.
Fipronil, human health concerns, **176**:53
Fipronil hydrolysis, **175**:160
Fipronil, hydrolysis, **176**:15
Fipronil, hydrolysis degradation rates, **176**:16
Fipronil, hydrolytic profile, **175**:232
Fipronil, insect cross-resistance, **176**:3, 52
Fipronil, insects controlled, **176**:8, 9
Fipronil, IPM compatibility, **176**:38, 52
Fipronil, leaching, adsorption, desorption, **176**:16
Fipronil, leaf residues over time, **176**:21, 22
Fipronil, LOCUSTOX study, **176**:29, 30, 38, 40
Fipronil, LOEL, **176**:36
Fipronil, major degradates common/chemical names, **176**:14
Fipronil, major degradates with structures, **176**:11
Fipronil, maximum residue levels in foodstuffs, **176**:20
Fipronil, metabolic pathway in rat, **176**:18
Fipronil, metabolism rates, **176**:33
Fipronil, microbial degradation half-lives, **176**:12
Fipronil, mode of action, **176**:3
Fipronil, mutagenic effects, **176**:36
Fipronil, neurotoxicity, **176**:34
Fipronil, NOEL, **176**:36
Fipronil, noninsecticidal effects agriculture, **176**:42
Fipronil, persistence/mobility in soil, **176**:16
Fipronil, photodegradation, **176**:12
Fipronil, physicochemical properties, **176**:3, 6
Fipronil, proprietary names, **176**:6, 8
Fipronil, reproductive effects, **176**:36
Fipronil, residue analysis, **176**:21

Fipronil, residues in drinking water, **176**:19
Fipronil, residues in food, **176**:19
Fipronil, risk assessments, **176**:52
Fipronil, soil residues over time, **176**:23
Fipronil, solubility different solvents, **176**:7
Fipronil, subchronic neurotoxicity, **176**:35
Fipronil, subchronic toxicity, **176**:35
Fipronil sulfide, **176**:14
Fipronil sulfone, **176**:14
Fipronil, termiticide, **176**:8
Fipronil toxicity, birds, **176**:32
Fipronil toxicity comparison, with fipronil-desulfinyl, **176**:37
Fipronil, usage/application rates, **176**:7, 9
Fipronil, veterinary risks, **176**:43
Fipronil, water half-life, **176**:12
Fish, arsenic effects, **180**:144
Fish bone properties, xenobiotic effects, **172**:1 ff.
Fish, ChE recovery times OP exposure, **172**:25
Fish contamination, veterinary medicines, **180**:24
Fish, monitors of PAH-contaminated sediment, **174**:12
Fish, pesticide residues, Gulf of Gdańsk, **179**:27
Fish skeletal system, organochlorine effects, **172**:15
Fish skeletal system, organophosphate effects, **172**:14
Fish, xenobiotic effects on skeletal system, **172**:1 ff.
Flame retardants, house dust contaminants, **175**:27
Flamprop-isopropyl, hydrolytic profile, **175**:194
Flamprop-methyl, hydrolytic profile, **175**:194
Fluazifop-butyl, hydrolytic profile, **175**:190
Flucythrinate, hydrolytic profile, **175**:192
Flumequine, environmental contamination, **180**:28
Flumethrin, hydrolytic profile, **175**:192
Flumichlorac pentyl, hydrolytic profile, **175**:204

Fluoranthrene (PAH), from coke production, **174:**13
Flupyrsulfuron methyl, hydrolytic profile, **175:**224
Flupyrsulfuron methyl, sulfonylurea bridge contraction, **175:**152
Fluroapatite, phosphate fertilizer, **177:**11
Fluroxypyr, hydrolytic profile, **175:**190
Fluvalinate, hydrolytic profile, **175:**192
Folpet, hydrolytic profile, **175:**204
Fonofos, hydrolytic profile, **175:**220
Food chain, metals accumulation from waste disposal on soils, **177:**2
Formamidine insecticides, hydrolysis, **175:**158
Formulations, fipronil insecticide, **176:**7
Fosmethilan photoproducts, **172:**205
Frazil ice, **171:**60
Freshwater contamination, veterinary medicines, **180:**18
Freundlich adsorption data, pyrethroids, **174:**56
Frog embryo teratogenesis assay–*Xenopus* (FETAX), **172:**35
Fumigants, mass transfer coefficients across plastic films, **177:**96, 101
Fundulus heteroclitus, acellular bone, **172:**10
Fungi, chromium toxicity, **178:**127

GA (tabun), review summary, **172:**71, 76
GA (tabun), RfDs & UFs, **172:**68
GABA-gated chloride currents, spinosad mode of action, **179:**40
GABA-regulated chloride channel, mode of action, **176:**3
Gafsa phosphate rock, fertilizer, **177:**20
Gallotia galloti (lizard), good OP exposure bioindicator, **172:**51
Garlon® (triclopyr), **174:**20
Gas chromatography detectors, methyl bromide, **177:**55
Gas chromatography, house dust contaminant analysis, **175:**14
Gas-phase photolysis, organophosphate photochemistry, **172:**199
GB (sarin), review summary, **172:**72, 76
GB (sarin), RfDs, & UFs, **172:**68
GD (soman), review summary, **172:**72, 76

GD (soman), RfDs & UFs, **172:**68
Genotoxicity, azaarenes, **173:**62
Genotoxicity, DINP, **172:**107
Genotoxicity, fipronil, **176:**36
Gentamicin, mechanisms of action, **171:** 39
GIFAP, (International Group of Natl Assoc of Mnftrs of Agrochem Prods), **177:**131
Glycocalyx, "slime", **173:**137
Gold miners, arsenic-caused diseases, **180:**136
Gold mining, arsenic discharges, **180:**133
Gold mining, arsenic soil contamination, **180:**141
Granulomatous amoebic encephalitis, **180:**104, 111
Grazon® (triclopyr), **174:**20
Ground state, photon absorption photochemistry, defined, **172:**131
Groundwater contamination, PFAS, PFOS, PFOA, **179:**109
Groundwater contamination, veterinary medicines, **180:**20, 30
Groundwater, triclopyr contamination, **174:**38
Gulf of Gdańsk, annual pollutant load, **179:**6
Gulf of Gdańsk, annual precipitation, **179:**8
Gulf of Gdańsk, biomass annual fluctuations, **179:**28
Gulf of Gdańsk, cadmium concentrations, **179:**17
Gulf of Gdańsk, coastal zone climate, **179:**7
Gulf of Gdańsk, copper concentrations, **179:**16
Gulf of Gdańsk, dominant algae groups, **179:**24
Gulf of Gdańsk, ecosystem changes, **179:** 1, 4
Gulf of Gdańsk, eutrophication, **179:**1 ff.
Gulf of Gdańsk, heavy metal pollutants, **179:**14, 18, 19
Gulf of Gdańsk, important ecosystem changes, **179:**7
Gulf of Gdańsk, lead concentrations, **179:** 16

Gulf of Gdańsk, map, **179**:5
Gulf of Gdańsk, morphological subunits, **179**:4
Gulf of Gdańsk, most polluted area of Baltic, **179**:2
Gulf of Gdańsk, nitrate/phosphate fluctuations, **179**:12
Gulf of Gdańsk, PAH annual fluctuations, **179**:29
Gulf of Gdańsk, PCB sediment levels, **179**:22
Gulf of Gdańsk, persistent organic pollutants, **179**:17
Gulf of Gdańsk, pesticide pollutants, **179**: 19, 24, 26
Gulf of Gdańsk, pesticides in fish, **179**:27
Gulf of Gdańsk, physical characteristics, **179**:4
Gulf of Gdańsk, phytoplankton pigment changes, **179**:23, 31
Gulf of Gdańsk, phytoplankton species, **179**:26
Gulf of Gdańsk, seasonal nutrient fluctuations, **179**:12, 14
Gulf of Gdańsk, seasonal oxygen changes, **179**:11
Gulf of Gdańsk, silicate fluctuations, **179**: 13, 16
Gulf of Gdańsk, surface chlorophyll seasonal changes, **179**:31
Gulf of Gdańsk vs Baltic Sea, comparisons, **179**:6
Gulf of Gdańsk, water temperature, **179**:8
Gulf of Gdańsk, zinc concentrations, **179**: 17

Hair color, effect trace element contamination, **175**:50
Hair, human, trace element concentrations, **175**:47 ff.
Hair location, effect trace element contamination, **175**:50
Hair, trace element contamination, see Human hair, **175**:47
Half-life, methyl bromide, **177**:59, 60, 61
Half-life, methyl bromide in soil, **177**:104
Half-life, PFOA/PFOS in humans, **179**: 114
Half-life, triclopyr in plants, **174**:27

Half-life, triclopyr in soils, **174**:23
Half-lives, fipronil, **176**:12
Half-lives, pesticides aquatic environment, **175**:190–233
Half-lives, pesticides in soil, **177**:127
Half-lives, pyrethroids in aerobic aquatic systems, **174**:66
Half-lives, pyrethroids in aerobic soil, **174**:64
Half-lives, pyrethroids in anaerobic water, **174**:65
Half-lives, triclopyr in waters, **174**:42
Haloxyfop ethotyl, hydrolytic profile, **175**:190
Hartmanella, previous genus name of *Acanthamoeba*, **180**:95
Hazard assessment, house dust contaminants, **175**:7
α-HCH (hexachlorocyclohexane), chiral insecticide, **173**:90, 93, 95
α-HCH, enantiomer structures, **173**:92
γ-HCH (lindane), chiral insecticide, **173**: 93
HCH (technical), world emission, **173**:93
HCH, CAS numbers, **173**:107
HCH, IUPAC chemical name, **173**:107
HCH-α (hexachlorocyclohexane), chiral insecticide, **173**:90, 93, 95
HCH-γ (hexachlorocyclohexane) lindane, chiral insecticide, **173**:93
HD (sulfur mustard), review summary, **172**:74, 78
HD (sulfur mustard), RfDs & UFs, **172**:68
Health effect, hair trace element contamination, **175**:55
Heavy metal adsorption, phosphate-induced, **177**:22
Heavy metal availability in soil-plant system, phosphorus role, **177**:1 ff.
Heavy metal immobilization in soils, phosphorus role, **177**:1 ff.
Heavy metal mobilization in soils, phosphorus role, **177**:1 ff.
Heavy metal pollutants, Gulf of Gdańsk, **179**:14, 18, 19
Heavy metal precipitation in soils, phosphate-induced, **177**:24
Heavy metal precipitation, phosphate liming action, **177**:27

House dust, indoor pollutants, **175:**1 ff.
House dust, ingestion with food, **175:**6
House dust, lead-contaminated, **175:**8
House dust, mean surface deposits, **175:**4
House dust, organophosphate pesticide contamination, **175:**8, 16
House dust, organotin concentrations, **175:**27
House dust, particle-size distribution, **175:**5
House dust, PCDF contaminants, **175:**27
House dust, pollutant sources, **175:**15
House dust, pollutants contained, **175:**1 ff.
House dust, quantity inhaled, **175:**5
House dust, sample preparation, **175:**13
House dust, sampling methods, **175:**9
House dust, seasonal deposits, **175:**4
House dust, skin absorption, **175:**6
House dust, suspended air concentrations, **175:**6
House dust, trace element contaminants, **175:**28, 30
House dust, xenobiotic benchmark concentrations, **175:**33
HT, RfDs & UFs, **172:**68
Human adverse reactions, fipronil, **176:**44
Human exposure, diisononyl phthalate, **172:**111
Human hair, age effect trace element contamination, **175:**52
Human hair, air pollution effect trace element contamination, **175:**61
Human hair, analytical methods trace elements, **175:**63, 65
Human hair, as trace element exposure indicator, **175:**49
Human hair, biological factors, **175:**49
Human hair, cosmetic use effect trace element contamination, **175:**59
Human hair, diet effect trace element contamination, **175:**54
Human hair, digestion trace element analysis, **175:**64
Human hair, environmental effects trace element contamination, **175:**61
Human hair, health effect trace element contamination, **175:**55
Human hair, nationality effect trace element contamination, **175:**60

Human hair, occupation effect trace element contamination, **175:**57
Human hair, race effect trace element contamination, **175:**60
Human hair, sample preparation trace element contamination, **175:**63
Human hair, sex effect trace element contamination, **175:**53
Human hair, trace element concentrations, **175:**47 ff.
Human hair, trace element contamination sources, **175:**48
Human hair, trace element exposure indicator, **175:**49
Human hair, trace element, location on body, **175:**50
Human health, arsenic effects, **180:**152
Human health, proposed arsenic criteria, **180:**155
Human poisoning, arsenic, **180:**154
Human toxicity, methyl bromide, **177:**50
Human toxicity, PFOA, PFOS, **179:**114
Humic substances, photosensitizing effects water, **172:**197
Hydrogen sulfide, Gulf of Gdańsk levels, **179:**4
Hydrolysis, bifenthrin, **174:**71
Hydrolysis, cyfluthrin, **174:**77, 79
Hydrolysis, cypermethrin, **174:**87, 90
Hydrolysis, deltamethrin, **174:**104, 108
Hydrolysis, esfenvalerate, **174:**121
Hydrolysis, fenpropathrin, **174:**134
Hydrolysis half-lives, pyrethroids, **174:**52
Hydrolysis, lambda-cyhalothrin, **174:**142
Hydrolysis, methyl bromide in water, **177:**59
Hydrolysis, permethrin, **174:**154, 160
Hydrolysis, pesticide waste disposal, **177:**150
Hydrolysis, pyrethroids, **174:**60
Hydrolysis, tralomethrin, **174:**109
Hydrolysis, triclopyr products, **174:**21
Hydroxy apatite, phosphate fertilizer, **177:**11
Hydroxyapatite, bone component, **172:**11
Hypochlorous acid, catalyzed by myeloperoxidase, **171:**4
Hypochlorous acid, chlorine gas added to water, **171:**4

Heavy metals, adsorption/desorption onto phosphates, **177**:21
Heavy metals, adsorption in soils, **177**:6
Heavy metals, complexation in soils, **177**:7
Heavy metals, defined, **177**:1
Heavy metals, fractionation categories in soils, **177**:8
Heavy metals, house dust contaminants, **175**:28, 30
Heavy metals in soils, anthropogenic processes, **177**:4
Heavy metals in soils, pedogenic processes, **177**:4
Heavy metals in soils, sources (table), **177**:3
Heavy metals, mode of action, **173**:8
Heavy metals, precipitation in soils, **177**:8
Heavy metals, reactions in soils (diagram), **177**:5
Heavy metals, soil immobilization by phosphate compounds, **177**:12, 13
Heavy metals, soil immobilization by water-soluble/insoluble phosphates, **177**:16
Heavy metals, soil mobilization by phosphate compounds, **177**:12, 13
Heavy metals, soil mobilization by water-soluble/insoluble phosphates, **177**:14
Heavy metals, solid-phase speciation, **177**:8
Heavy metals, sources in soil environment, **177**:4
Helsinki Commission Baltic Monitoring Programme, **179**:2
Henry's law constant, methyl bromide, **177**:68, 69
Henry's law constants, pyrethroids, **174**:52, 54
Henry's law constants, soil fumigants, **177**:107
Heptachlor *exo*-epoxide, CAS number, **173**:107
Heptachlor *exo*-epoxide, chiral insecticide, **173**:90, 99, 104
Heptachlor *exo*-epoxide, enantiomer structures, **173**:92
Heptachlor *exo*-epoxide, IUPAC chemical name, **173**:107

Heptachlor, hydrolytic profile, **17**:
Heraeus Suntest, photochemical r described, **172**:145
Herbicides, aryloxyalkanoate hyd **175**:91
Herbicides, leaching indices, **174**
Hexachlorobenzene, vertebrate e **176**:74
Hexachlorocyclohexane (α-HCH insecticide, **173**:90, 93, 95
Hexachlorocyclohexane (HCH), reoisomers, **173**:93
Hexachlorocyclohexane (HCH), mesoforms, **173**:93
High volume small surface sam house dust, **175**:9
HN2, RfDs & UFs, **172**:68
HOMO-LUMO gap energies, P 59
Honeybee larvae, spinosad toxi 53, 59
Honeybee pesticide toxicity gr defined, **179**:66
Honeybee, spinosad foraging t **179**:45
Honeybee, spinosad toxicity, **1**
Horizontal flux method, methy volatilization, **177**:88
Hormidium rivulare, Cu toxici **178**:29
House dust, air concentration **175**:9
House dust, as exposure mark
House dust, composition & p **175**:3
House dust, contaminant anal ods, **175**:13
House dust, contaminants ha ment, **175**:7
House dust, dioxin contamin:
House dust, endocrine-disrup contaminants, **175**:26
House dust, EPA contamina **175**:32
House dust, EPA definition,
House dust, flame retardant **175**:27
House dust, heavy metal co **175**:28, 30

Imidacloprid hydrolysis, **175**:159
Imidacloprid, hydrolytic profile, **175**:232
Imiprothrin, hydrolytic profile, **175**:192
Impurities in electrochemical production of POSF, **179**:101
Incineration, pesticide contaminated soils, **177**:163
Incineration, pesticide disposal method, **177**:136
Incineration, pesticide plastic & paper packaging, **177**:143
Indeno[1,2,3-*cd*]pyrene, PAH, **179**:75
Indoor air, house dust pollution, **175**:1 ff.
Indoor pollutants, boiling point range (table), **175**:2
Indoor pollutants, classification (table), **175**:2
Indoor pollutants, house dust, **175**:1 ff.
Indoor residence times, **175**:1 ff.
Inorganic metals, tissue residues, **173**:23
Insect resistance, fipronil, **176**:3, 52
Insecticide formulations, fipronil, **176**:7
Insecticides, house dust quantities, **175**:19
Integrated pest management (IPM), fipronil compatibility, **176**:38
International Group of Natl Assoc of Mnftrs of Agrochem Prods (GIFAP), **177**:131
International Union of Pure and Applied Chemistry, (IUPAC), **177**:125
Interspecific differences, algal toxicant sensitivity, **178**:35
Intraspecies variability, algal toxicant sensitivity, **178**:36
Invertebrates, metal bioavailability in soils, **178**:10, 11
Invertebrates, PAH toxicity, **179**:78
Iodine number, fruit cuticle parathion photochemical half-life, **172**:165
Iodofenphos photoproducts, **172**:176
IPM, fipronil compatibility, **176**:38, 52
Iprodione, hydrolytic profile, **175**:204
Iron concentrations, Antarctic fish, **171**: 53 ff., **171**:63
Iron, effects on chromium oxidation, **178**: 98
Iron fixation in soils, **178**:6
Iron, occupational hair levels, **175**:58
Iron transformation in soils, **178**:4

Irradiation methods, photochemistry, **172**: 144
Irreversible reduction, Cr(VI) to Cr(III), **178**:78
ISO, International Organization for Standardization, **178**:34
Isofenphos, hydrolytic profile, **175**:220
Isofenphos photoproducts, **172**:214
Isoniazid, mechanisms of action, **171**:39
Isoquinoline, microbial degradation pathway, **173**:44
Isoquinoline, toxicity, **173**:42
Isoxaben, acid hydrolysis pathway, **175**: 104
Isoxaben, hydrolytic profile, **175**:196
Isoxaflutole, hydrolysis pathway, **175**:162
Isoxaflutole, hydrolytic profile, **175**:232
IUPAC (International Union of Pure and Applied Chemistry), **177**:125
IUPAC chemical names, chiral pesticides, **173**:107
Ivermectin (22,223-dihydroavermectin B_1), synthetic avermectin B_1, **171**: 112
Ivermectin, binding in aquatic systems, **171**:115
Ivermectin, dung fauna effects, **171**:117
Ivermectin, environmental contamination, **180**:28
Ivermectin, livestock pests effects, **171**: 117
Ivermectin, soil/dung concentrations, **171**: 114
Ivermectin, soil half-life, **171**:114, 118

K-Othrin®, deltamethrin, **174**:95
Kanamycin resistant bacteria, **171**:21, 23
K_d (adsorption coefficient), methyl bromide, **177**:68
K_d (Partition coefficients), metals in soils, **178**:15
Kepone-induced fish skeletal anomalies, **172**:13
K_h, pyrethroids, **174**:52, 54
Kinetic analyses, pyrethroids, **174**:58
Kinetics, pesticide hydrolysis, **175**:81
Kiwifruit flowers, spinosad honeybee toxicity, **179**:47
K_{oc}, pyrethroids, **174**:52

K_{ow}, azaarenes related to baseline toxicity, **173**:48
K_{ow}, pyrethroids, **174**:52
K_{ow}, pyrethroids, **174**:52
KPEG, see potassium polyethylene glycol ether, **177**:140
Kresoxim-methyl, hydrolytic profile, **175**:232
Krill, high fluorine content, **171**:62
K_{rs}, pyrethroids, **174**:55

L (Lewisite), review summary, **172**:75, 78
Lambda-cyhalothrin, abiotic chemical properties, **174**:142
Lambda-cyhalothrin, aerobic aquatic degradation, **174**:149, 150
Lambda-cyhalothrin, aerobic soil degradation, **174**:148, 149
Lambda-cyhalothrin, biotic chemical properties, **174**:148
Lambda-cyhalothrin, hydrolysis, **174**:142
Lambda-cyhalothrin, photolysis in water, **174**:142, 147
Lambda-cyhalothrin, photolysis on soil, **174**:146, 148
Lambda-cyhalothrin, physicochemical properties, **174**:52, 139
Lambda-cyhalothrin, soil sorption, **174**:143, 145
Landfarming, pesticide contaminated soil cleanup, **177**:169
Lead concentrations, Antarctic fish, **171**:53 ff., 63
Lead, EPA house dust contaminant guidelines, **175**:32
Lead fixation in soils, **178**:5
Lead, house dust concentrations, **175**:29
Lead, occupational hair levels, **175**:58
Lead shot, vertebrate biomonitor effects, **176**:103, 105, 106, 109, 111
Lead, soil contaminant sources, **177**:3
Lead-based paint, house dust lead source, **175**:16
Lead-phosphate interactions, soils, **177**:26
Leaded gasoline, atmospheric methyl bromide source, **177**:53
Leather tanning with chromium, described, **178**:54

Legionella pneumophila, *Acanthamoeba* interactions, **180**:114
Legionellae survival, different lab conditions, **171**:16
Leptophos, hydrolytic profile, **175**:220
Lewisite (L), review summary, **172**:75, 78
Lewisite, environmental fate, **172**:79
Lewisite, RfDs & UFs, **172**:68
Lichens, Antarctic biomonitors, **171**:55
Lichens, baseline trace metal levels Antarctica, **171**:87, 92
Light, effect in algal toxicity testing, **178**:33
Light sources, photochemical reactions, **172**:141
Liming action of phosphate, heavy metal precipitation, **177**:27
Liming, increases metal retention in soils, **177**:8
Lindane (γ-HCH), chiral insecticide, **173**:93
Lindane, house dust concentrations, **175**:19
Linuron, hydrolytic profile, **175**:198
Livestock medicines, environmental pathways, **180**:14
Lizard *(Gallotia galloti)*, good OP exposure bioindicator, **172**:51
Lizards, parathion-exposed BChE/CbE activity, **172**:43
LOCUSTOX study, fipronil, **176**:29, 30, 38, 40
$LOEC_{50}$ values, PFAS, PFOS, aquatic organisms, **179**:112
LOEL, fipronil, **176**:36
LOELs, maternal/developmental DINP, **172**:104

Malaoxon I_{50}s, ChE fish, **172**:28
Malathion, hydrolysis pathways, **175**:139
Malathion I_{50}s, ChE fish, **172**:28
Malathion photoproducts, **172**:200
Mammalian toxicity, avermectins, **171**:126
Manganese (II) oxidation to Mn(IV), microbial, **178**:122
Manganese (II) oxidizing microbes, list, **178**:123

Manganese concentrations, Antarctic fish, **171**:53 ff., 63
Manganese effects on chromium oxidation, **178**:99
Manganese fixation in soils, **178**:6
Manganese, house dust concentrations, **175**:29
Manganese, occupational hair levels, **175**:58
Manganese oxidation, microbial mechanism, **178**:123
Manganese oxides formed by microbes, **178**:124
Manganese, soil contaminant sources, **177**:3
MAR (multiple-antibiotic-resistant) bacteria, **171**:19
MAR-bacteria selection, wastewater chlorination, **171**:22
Marine fish, arsenic effects, **180**:145
Marine invertebrates, arsenic effects, **180**:145
Marine mammals, mercury bioaccumulation, **171**:69
Marine mammals, trace metal concentrations, **171**:72
Marine phytoplankton pigment shifts, **179**:23, 31
Marine plants, arsenic effects, **180**:144
Marine turtles, xenobiotic bioindicators, **172**:34
Mass spectrometry, house dust contaminant analysis, **175**:14
Maximum residue levels foodstuffs, fipronil, **176**:20
MDPA (chlorodihydroxy pyridinyloxyacetic acid), triclopyr photoproduct, **174**:24
MeBr (methyl bromide), **177**:46
Mechanisms of action, antibiotics in bacteria, **171**:39
Mechanisms of damage to bacteria, UV light, **171**:29
Mechanisms of resistance (bacterial) to antibiotics, **171**:40
Mecoprop, CAS number, **173**:107
Mecoprop, chiral herbicide, **173**:90, 94, 96
Mecoprop, enantiomer structures, **173**:92

Mecoprop, IUPAC chemical name, **173**:107
Medicines veterinary, environmental contaminants, **180**:1 ff.
Membrane permeability, chlorine effects, **171**:6
Mephospholan photoproducts, **172**:213
Mercury arc, high-pressure bayonet, described, **172**:145, 146
Mercury, bioaccumulation in pelagic seabirds/marine mammals, **171**:69
Mercury, concentrations Antarctic fish, **171**:53 ff., 63
Mercury, feather concentrations seabirds Southern/Northern hemispheres, **171**:70
Mercury, house dust concentrations, **175**:30
Mercury, human hair concentrations, **175**:59
Mercury pollution, Minamata Bay (Japan), **174**:9
Mercury, soil contaminant sources, **177**:3
Mercury, transfer to birds/seals breeding in Antarctica, **171**:77
Mercury, vertebrate biomonitor effects, **176**:96, 98, 101, 102
Metabolic pathway, fenvalerate in animals, **176**:150
Metabolic pathway, fipronil in rat, **176**:18
Metal aging in soils, **178**:1 ff.
Metal aging, natural soils, **178**:6
Metal availability in soils, aging effects, **178**:1 ff.
Metal cations, complexation affinity by organic matter (diagram), **177**:7
Metal deposition biomonitoring, Antarctic stations, **171**:91
Metal extraction from soils, **178**:6
Metal phosphate compounds, water solubility, **177**:25
Metal phosphate precipitation, primary P-induced immobilization, **177**:8
Metal phosphates, soil precipitates, **177**:24
Metal sorption/desorption, soils, **178**:3
Metal-contaminated soils, risk assessment, **178**:1 ff.
Metal-organic complex formation, factors affecting, **177**:7

Metalaxyl, hydrolytic profile, **175**:194
Metalloids, defined, **177**:2
Metals, environmental availability, defined, **178**:1
Metals, environmental bioavailability, defined, **178**:1
Metals, toxicological bioavailability, defined, **178**:1
Metam sodium, methyl bromide replacement, **177**:107
Methamidophos, hydrolysis pathways, **175**:140
Methane-consuming bacteria, rusticle component, **173**:135
Methidathion photoproducts, **172**:206
Methidathion, sublethal AChE inhibition aquatic organisms, **172**:47
Methodology, enantiomeric enrichment, **173**:88
Methomyl, hydrolytic profile, **175**:200
Methoxychlor, house dust concentrations, **175**:19
Methoxychlor, hydrolysis pathways, **175**:126
Methoxychlor, hydrolytic profile, **175**:206
Methyl bromide (MeBr), **177**:46
Methyl bromide, adsorption coefficient, **177**:68
Methyl bromide, air sampling & analysis, **177**:54, 55
Methyl bromide, air sampling adsorbents, **177**:56
Methyl bromide, air sampling containers, **177**:56
Methyl bromide, application depth effect, **177**:100
Methyl bromide, atmospheric oceanic emissions, **177**:53
Methyl bromide, bacterial biodegradation, **177**:61
Methyl bromide, chemical alternatives, **177**:47
Methyl bromide, chloropicrin warning agent, **177**:49
Methyl bromide, containment to minimize volatilization, **177**:94
Methyl bromide, degradation rate effect on emissions, **177**:104
Methyl bromide, degraded to bromine ion in soil, **177**:82

Methyl bromide, developing alternatives, **177**:106
Methyl bromide, diffusion in air, **177**:78
Methyl bromide, diffusion in soils, **177**:77
Methyl bromide, effective soil diffusion coefficient, **177**:75
Methyl bromide, effects of organic matter on hydrolysis, **177**:60
Methyl bromide, elimination economic impact, **177**:47
Methyl bromide, enhancing soil transformation rates, **177**:62
Methyl bromide, environmental concerns, **177**:46
Methyl bromide, environmental fate, **177**:45 ff.
Methyl bromide, environmental transformation, **177**:58
Methyl bromide, estimating total loss from soil, **177**:81
Methyl bromide, fumigation application methods, **177**:51
Methyl bromide, fumigation Br residues in soils/plants, **177**:62, 64
Methyl bromide, fumigation soil bulk density effect, **177**:103
Methyl bromide, fumigation soil water content effect, **177**:102
Methyl bromide, fumigation tarp coverings, **177**:51
Methyl bromide, fumigation uses, **177**:48, 50
Methyl bromide, gas chromatography detectors, **177**:55
Methyl bromide, global usage (budget), **177**:47
Methyl bromide, human toxicity, **177**:50
Methyl bromide, hydrolysis in soil, **177**:60
Methyl bromide, hydrolysis water effects, **177**:59
Methyl bromide, injection depth & use of plastic film effects, **177**:99, 101
Methyl bromide, known atmospheric sources, **177**:53
Methyl bromide, methods for minimizing volatilization, **177**:94
Methyl bromide, mobility indices, **177**:73

Methyl bromide, natural atmospheric removal, **177**:53
Methyl bromide, ozone depletion potential, **177**:46, 52
Methyl bromide, phase partitioning, **177**:68
Methyl bromide phase-out, economic concerns, **177**:47
Methyl bromide phase-out, major crop loss estimates, **177**:48
Methyl bromide phase-out, Montreal Protocol agreement, **177**:50
Methyl bromide, photohydrolysis, **177**:60
Methyl bromide, physical/chemical properties, **177**:49
Methyl bromide, phytotoxicity, **177**:66
Methyl bromide, plastic film barriers, **177**:95
Methyl bromide, production by growing plants, **177**:67
Methyl bromide, reactions with ozone, **177**:52
Methyl bromide, replacement fumigants, **177**:107
Methyl bromide, simulated environmental fate, **177**:70
Methyl bromide, soil fumigant history, **177**:50
Methyl bromide, soil gas distribution after injection, **177**:78
Methyl bromide, soil gas partitioning, **177**:79
Methyl bromide, soil half-life, **177**:104
Methyl bromide, solubility vs temperature, **177**:69
Methyl bromide, transport in dry soils, **177**:74
Methyl bromide, transport model, **177**:70
Methyl bromide, vapor pressure vs temperature, **177**:69
Methyl bromide, volatilization aerodynamic method, **177**:85
Methyl bromide, volatilization boundary condition, **177**:71
Methyl bromide, volatilization chamber methods, **177**:83
Methyl bromide, volatilization experiments & mass balance, **177**:90
Methyl bromide, volatilization field experiments, **177**:89
Methyl bromide, volatilization from soil, **177**:81
Methyl bromide, volatilization horizontal flux method, **177**:88
Methyl bromide, volatilization micrometeorological methods, **177**:85
Methyl bromide, volatilization theoretical profile shape method, **177**:86
Methyl bromide, volatilization trajectory simulation, **177**:87
Methyl isothiocyanate (MITC), methyl bromide replacement, **177**:107
Methyl mercury, mercury source in Antarctic seabirds/marine mammals, **171**:69
Methyl paraoxon I_{50}s, ChE fish, **172**:28
Methyl parathion, hydrolytic profile, **175**:208
Methyl parathion, sublethal AChE inhibition aquatic organisms, **172**:46
Methylene blue, photooxidation organophosphates, **172**:174
Methylococcus sp., methyl bromide biodegradation, **177**:61
Methylomonas sp., methyl bromide biodegradation, **177**:61
Metolachlor, acid hydrolysis pathway, **175**:102
Metolachlor, hydrolytic profile, **175**:194
Metsulfuron methyl, hydrolytic profile, **175**:224
Mevinphos, hydrolysis pathways, **175**:135
Mevinphos, hydrolytic profile, **175**:212
Mexacarbate, hydrolytic profile, **175**:200
MFO (mixed-function oxidase), amphibians/reptiles, **172**:35
Micelles, abiotic pesticide hydrolysis, **175**:89
Microbial consortial events, deep sea rusticles, **173**:117 ff.
Microbial consortium, defined, **173**:122
Microbial degradation, azaarenes more than three aromatic rings, **173**:46
Microbial degradation pathway, isoquinoline, **173**:44
Microbial degradation pathway, quinoline, **173**:43
Microbial oxidation, Mn(II) to Mn(IV), **178**:122
Microbial reduction, Cr(VI), **178**:106

Microcystis aeruginosa (alga), toxicity testing, **178:**34
Micrometeorological methods, methyl bromide volatilization, **177:**85
Microorganisms, aging metal soil bioavailability, **178:**13
Microorganisms, capable of oxidizing manganese, **178:**123
Microorganisms, capable of reducing Cr(VI), list, **178:**106
Microorganisms, PAH toxicity, **179:**76
Microorganisms, tolerance to Cr(VI), **178:**103
Milk, aflatoxins FDA action levels, **171:**144
Minamata Bay (Japan), mercury pollution, **174:**9
Mirror image structures, chiral compounds, **173:**92
Mixed-function oxidase (MFO), amphibians/reptiles, **172:**35
γ-MnOOH, oxidation rate of Cr(III), **178:**102
Mobam, hydrolytic profile, **175:**200
Mobility indices, methyl bromide, **177:**73
Mode of action, avermectins, **171:**112
Mode of action, DINP, **172:**109
Mode of action, fipronil, **176:**3
Mode of action, organophosphate insecticides, **172:**150
Mode of action, spinosad insects, **179:**39
Mode of action, triclopyr herbicide, **174:**20
Modes of action, chemical classes, **173:**5, 8
Modes of action, narcotic chemicals, **173:**3, 5
Molecular weights, pyrethroids, **174:**52
Mollusks, trace metal monitors Southern Ocean, **171:**74, 75
Molybdenum, house dust concentrations, **175:**30
Molybdenum, soil contaminant sources, **177:**3
Monocrotophos, hydrolytic profile, **175:**212
Montreal Protocol, methyl bromide phase-out agreement, **177:**50
Monuron, hydrolytic profile, **175:**198

Mosses, biomonitors atmospheric metals, **171:**90, 92
Mosses, elemental composition Antartica, **171:**89, 92
Multihit disinfectant, chlorine, **171:**5
Multiple-antibiotic-resistant (MAR) bacteria, **171:**19
Mussel Watch program, described, **172:**23
Mussels, as bioindicators water pollution, **172:**23
Mutagenic potentials, aflatoxin-contaminated feeds, **171:**167
Mutagenicity, benzacridines, **173:**65
Mutagenicity, benzoquinolines & metabolites, **173:**64, 65
Mutagenicity, dibenzacridines, **173:**66
Mutagenicity, DINP, **172:**107
Mutagenicity, fipronil, **176:**36
Mutagenicity, quinoline, **173:**63
Mutation assay methods, DINP, **172:**107
Mutatox™, azaarene genotoxicity tests, **173:**62, 64
Mycorrhizal fungi, chromium toxicity, **178:**127
Mycorrhizal rhizosphere, effects on metal tolerance in plants, **177:**31
Mycorrihizal fungi colonization of plant roots, rhizosphere microflora, **177:**33
Myeloperoxidase, catalysis to hypochlorous acid, **171:**4

N-heterocyclic PAHs, **173:**40
Naled, hydrolytic profile, **175:**212
Naptalam, hydrolytic profile, **175:**196
Napthalene, PAH, **179:**75
Narcotic chemicals, **173:**7 ff.
Narcotic chemicals, modes of action, **173:**3
Narcotic chemicals, polar/nonpolar, **173:**3
Narcotics, CBRs, **173:**5
Narcotics, mode of action, **173:**8
National Registration Authority, Australia, **177:**129
National Research Council (NRC) Review, chem warfare agents, **172:**71
Nationality effect, hair trace element contam, **175:**60
Natural soils, metal aging, **178:**6
Navicula pelliculosa (alga), toxicity testing, **178:**34

Neburon, hydrolytic profile, **175**:198
Neuropathy target esterase, described, **172**:150
Neurotoxicity, fipronil, **176**:34, 35
Nickel concentrations, Antarctic fish, **171**:53 ff., 63
Nickel, occupational hair levels, **175**:58
Nickel, soil contaminant sources, **177**:3
Nickle fixation in soils, **178**:6
Niclosamide, hydrolytic profile, **175**:196
Nicosulfuron, hydrolytic profile, **175**:224
Nicotinic receptors, spinosad mode of action, **179**:40
Nitrification, Cr(III) inhibitory effects in soil, **178**:139
Nitrifying bacteria, methyl bromide biodegradation, **177**:61
Nitroaryl compounds, UV reaction to nitrosoaryl derivatives, **172**:162
Nitrosococcus sp., methyl bromide biodegradation, **177**:61
Nitrosolobus sp., methyl bromide biodegradation, **177**:61
Nitrosomonas sp., methyl bromide biodegradation, **177**:61
Nitszchia closterium (marine diatom), Cu & Zn tests, **178**:28
Nitzschia sp. (alga), toxicity tests, **178**:34
NOEL, fipronil, **176**:36
NOELs, maternal/developmental DINP, **172**:104
Non-target insects, avermectin effects, **171**:116
Nonessential elements, plant/animal nutrition defined, **177**:2
Nontarget toxicity, spinosad, **179**:40
Notothenioids, Antarctic biomonitors, **171**:54
NRC (National Research Council) Review, chemical warfare agents, **172**:71
NRC Recommendations, chemical warfare agents Army evaluation, **172**:76
Nuclear waste, steel drum deterioration on ocean floor, **173**:134
Nucleophiles, methyl bromide hydrolysis, **177**:59
Nutrients, effect in algal toxicity tests, **178**:31

Obesity effect, hair trace element contam, **175**:58
Obsolete pesticide stocks, IUPAC disposal recommendations, **177**:180
Obsolete pesticides, bulk disposal methods, **177**:135
Occupation effect, hair trace element contamination, **175**:55
Occupational exposure, diisononyl phthalate, **172**:111
Octanol-water partition coefficients (K_{ow}), pyrethroids, **174**:52, 53
OECD (See Organization for Economic Cooperation & Develop), **178**:8
OECD, Organization for Economic Cooperation & Develop, **178**:34
OECD standard artificial soils, **178**:8
OP acute toxicity in fish, reasons for differences, **172**:32
OP concentrations, ChE I_{50}s various wildlife (table), **172**:28
OP exposure, biomonitoring aquatic environments, **172**:42
OP photochemistry, **172**:129 ff.
OP potentiation by ergosterol biosynthesis inhibit (EBI) fungicides, birds, **172**:39
OP potentiation, carbamate (carbaryl), **172**:39
OPs (organophosphates), photochemistry, **172**:129 ff.
OPs, photochemical absorption coefficients (table), **172**:149
OPs, sublethal AChE inhibition aquatic organisms, **172**:47
Oral reference doses (RfDs), chemical warfare agents, **172**:65 ff.
Organic chemicals, abbreviations, **172**:119
Organic matter, effects on methyl bromide hydrolysis, **177**:60
Organochlorine insecticides, house dust concentrations, **175**:18
Organochlorine insecticides, hydrolysis mechanisms, **175**:124
Organochlorine insecticides, structures & hydrolytic profiles, **175**:206
Organochlorine insecticides, vertebrate effects, **176**:74

Organochlorine residues, fish, **179**:27
Organochlorine-induced effects in fish skeleton, **172**:15
Organometallic chemicals, tissue residues, **173**:25
Organometals, mode of action, **173**:8
Organophosphate insecticides, AChE inhibition, **172**:2
Organophosphate insecticides, CAS numbers, **172**:215
Organophosphate insecticides, chemical names, **172**:215
Organophosphate insecticides, common names, **172**:215
Organophosphate insecticides, history, **172**:148
Organophosphate insecticides, house dust contamination, **175**:8, 16
Organophosphate insecticides, mode of action, **172**:150
Organophosphate nerve agents, air exposure limits, **172**:66
Organophosphate pesticides, nervous system effects, **172**:2
Organophosphate-induced effects in fish skeleton, **172**:14
Organophosphates, hydrolytic half-lives, **172**:151
Organophosphates, photochemical absorption coefficients (table), **172**:149
Organophosphorus (OP) insecticides, wildlife exposure, **172**:21 ff.
Organophosphorus ester pesticides, hydrolysis mechanisms, **175**:127
Organophosphorus esters, acid and alkaline hydrolysis mechanisms, **175**:128
Organophosphorus insecticides, catalytic hydrolysis, **175**:182
Organophosphorus insecticides, history, **172**:148
Organophosphorus insecticides photochemistry, **172**:129 ff.
Organophosphorus insecticides, structures & hydrolytic profiles, **175**:208
Organotin compounds, house dust concentrations, **175**:27
Oxamic acid, triclopyr photoproduct, **174**:24
Oxamyl, hydrolytic profile, **175**:200

Oxazoline, alkaline hydrolysis pathway, **175**:106
Oxidation, chromium (III) to (VI), **178**:71
Oxidation, pesticide waste disposal, **177**:152
Oxime ether herbicides, hydrolysis kinetics mechanisms, **175**:153
Oxime ether herbicides, structures & hydrolytic profiles, **175**:228
Oxime N-alkylcarbamate hydrolysis, **175**:114
Oxolinic acid, wild fish contamination, **180**:26
Oxycarboxin, hydrolytic profile, **175**:196
Oxycarboxin, surface-catalyzed hydrolysis, **175**:169
Oxychlordane, CAS number, **173**:107
Oxychlordane, chiral insecticide, **173**:90, 100, 104
Oxychlordane, enantiomer structures, **173**:92
Oxychlordane, IUPAC chemical name, **173**:107
Oxydemeton-methyl, hydrolytic profile, **175**:212
Oxytetracycline, marine fish contamination, **180**:26
Ozonation, pesticide waste disposal, **177**:152, 154
Ozone "hole" discovery, Antarctica, **171**:54
Ozone Assessment Synthesis Panel (U.N.), **177**:46
Ozone breakdown efficiency, bromine vs chlorine, **177**:47
Ozone depletion, chlorinated compounds, **177**:46
Ozone depletion, methyl bromide, **177**:52
Ozone depletion potential index, **177**:46
Ozone depletion potential, methyl bromide, **177**:46

PAH (see also Polycyclic Aromatic Hydrocarbons), **179**:73 ff.
PAH, chemical reaction in water column (diagram), **173**:55
PAH contamination, coke production by-product, **174**:12

PAH ecotoxicity data for soil quality criteria, **179**:73 ff.
PAH soil concentrations, Denmark, **179**:89
PAH toxicity, species sensitivity distribution method, **179**:85
PAH-sediment remediation, Black River, **174**:12
PAHs (polycyclic aromatic hydrocarbons), **173**:40
PAHs (polycyclic aromatic hydrocarbons), **174**:12
PAHs, (Polycyclic Aromatic Hydrocarbons), ecotoxicity data, **179**:73 ff.
PAHs (see also Polycyclic Aromatic Hydrocarbons), **179**:73 ff.
PAHs, biotransformation and metabolism, **173**:41
PAHs, from coke production, **174**:13
PAHs, HOMO-LUMO gap energies, **173**:59
PAHs, house dust contaminants, **175**:13, 24
PAHs, levels Gulf of Gdańsk, **179**:29
PAHs, list of those tested, **179**:75
PAHs, phototoxic effects on aquatic organisms, **173**:57
PAHs, predicted environmental no effect concentration, **179**:83
PAHs, sediment removal effects, **174**:12
PAHs, soil quality criteria calculation, **179**:81
PAHs, soil quality standards, **179**:73 ff.
PAHs, terrestrial ecotoxicity data, **179**:74
PAHs, toxicity earthworms, **179**:75
PAHs, toxicity enchytraeids, **179**:75
PAHs, toxicity invertebrates, **179**:78
PAHs, toxicity microorganisms, **179**:76
PAHs, toxicity mustard, red clover, ryegrass, **179**:74
PAHs, toxicity plants, **179**:76
PAHs, toxicity soil invertebrates, **179**:74
PAHs, toxicity springtails, **179**:75
Paraoxon, formation reaction pathway, **172**:161
Paraoxon $I_{50}s$, ChE fish, **172**:28
Paraoxon $I_{50}s$, ChE frog, **172**:30
Parathion, catalytic hydrolysis, **175**:183
Parathion, hydrolytic profile, **175**:208
Parathion $I_{50}s$, ChE fish, **172**:28
Parathion methyl, hydrolytic profile, **175**:208
Parathion, paraoxon formation reaction pathway, **172**:161
Parathion photochemical half-life, fruit cuticle iodine number, **172**:165
Parathion, photodegradation different solvents, **172**:138
Parathion photoproducts, **172**:157, 159, 164
Particle-size distribution, house dust, **175**:5
Particulate organic matter (POM), **175**:2
Partition coefficients (K_d), metals in soils, **178**:15
Pathogenic protozoa, disease symptoms, **180**:94
PCBs (polychlorinated biphenyls), house dust contaminants, **175**:14, 24
PCBs, atropisomer of methylsulfonyl-PCBs, chiral PCBs, **173**:91
PCBs, carp Waukegan Harbor (graph), **174**:11
PCBs, chiral, **173**:91
PCBs, degraded by KPEG, **177**:141
PCBs, Lake Michigan fish, **174**:11
PCBs, levels Gulf of Gdańsk, **179**:19, 20, 21
PCBs, sediment levels Baltic Sea, **179**:22
PCBs, sediment removal effects, **174**:10
PCBs, tissue residues aquatic organisms, **173**:22
PCBs, vertebrate effects, **176**:74
PCDDs (dioxins), house dust contaminants, **175**:27
PCDDs, produced by chlorinated pesticide incineration, **177**:140
PCDDs, tissue residues aquatic organisms, **173**:22
PCDDs, vertebrate effects, **176**:74
PCDFs (polychlorinated furans), house dust contaminants, **175**:27
PCDFs, produced by chlorinated pesticide incineration, **177**:140
PCDFs, tissue residues aquatic organisms, **173**:22
PCDFs, vertebrate effects, **176**:74
Peanut meal, aflatoxin-contaminated rat feeding studies, **171**:163

Pedogenic processes, heavy metals in soils, **177**:4
Pelagic seabirds, mercury bioaccumulation, **171**:69
Penconazole, ergosterol biosynthesis inhibiting fungicides, **172**:39
Penicillins, mechanisms of action, **171**:39
β-pentachlorocyclohexene, chiral PCB, **173**:91
γ-pentachlorocyclohexene, chiral PCB, **173**:91
Pentachlorocyclohexene-β, chiral PCB, **173**:91
Pentachlorocyclohexene-γ, chiral PCB, **173**:91
Pentachlorophenol (PCP), house dust concentrations, **175**:19, 22
Perfluoroalkyl phosphate, chemical structure, **179**:103
Perfluoroalkylated substances (PFAS), environ fate/occurrence, **179**:102
Perfluoroalkylated substances (PFAS), environ/toxicity effects, **179**:99 ff.
Perfluoroalkylbetaine, fire-fighting foams, **179**:104
Perfluoroalkylethylates (PFAS), products of telomerization, **179**:101
Perfluoroalkylethylates, degradation products, **179**:99
Perfluoroalkylethylates, uses, **179**:100
Perfluoroalkylsulfonates, degradation products, **179**:99
Perfluoroalkylsulfonates, uses, **179**:100
Perfluorooctanoic acid (PFOA), groundwater occurrence, **179**:109
Perfluorooctanoic acid, structure, **179**:99, 100
Perfluorooctyl sulfonate (PFOS), groundwater occurrence, **179**:109
Perfluorooctyl sulfonate, structure, **179**:99, 100
Permethrin, abiotic chemical properties, **174**:154
Permethrin, aerobic soil degradation, **174**:157, 162
Permethrin, anaerobic soil degradation, **174**:157
Permethrin, biotic chemical properties, **174**:157

Permethrin, house dust concentrations, **175**:19
Permethrin, hydrolysis, **174**:154, 160
Permethrin, hydrolytic profile, **175**:192
Permethrin, photolysis in water, **174**:154, 160
Permethrin, photolysis on soil, **174**:154, 161
Permethrin, physicochemical properties, **174**:52, 151
Permethrin, soil sorption, **174**:153, 155, 158
Peroxisome proliferator, PFOA, **179**:111
Persistence, triclopyr in soils, **174**:40
Persistence, triclopyr in water/sediments, **174**:36
Persistent organic pollutants, utility index & scores, **176**:76
Persistent organic pollutants, vertebrate effects, **176**:74
Persistent organic pollutants, vertebrate vulnerability index, **176**:77
Persistent organic pollutants, vertebrates rank order index, **176**:79
Perylene, PAH, **179**:75
Pesticide 50% dissipation times ($DT_{50\%}$), in soil, **177**:127, 158
Pesticide abiotic hydrolysis, aquatic environment, **175**:79 ff.
Pesticide aquatic toxicity, veterinary use, **180**:44 ff.
Pesticide candidates for phytoremediation cleanup, **177**:171
Pesticide CAS numbers list, **177**:183
Pesticide CAS numbers, veterinary use (table), **180**:76
Pesticide chemical names list, **177**:183
Pesticide Chemical Abstracts Service (CAS) numbers list, **177**:183
Pesticide common names list, **177**:183
Pesticide concentrations in waste, user sites, **177**:126
Pesticide container & rinsewater disposal, IUPAC recommends, **177**:180
Pesticide container and packaging disposal, **177**:142
Pesticide container recycling, **177**:142
Pesticide contaminated soils, bioaugmentation, **177**:170

Pesticide contaminated soils, biological treatment, **177:**165
Pesticide contaminated soils, biostimulation, **177:**178
Pesticide contaminated soils, cleanup methods, **177:**161
Pesticide contaminated soils, composting, **177:**166
Pesticide contaminated soils, incineration, **177:**163
Pesticide contaminated soils, IUPAC cleanup recommends, **177:**180
Pesticide contaminated soils, KPEG treatment, **177:**164
Pesticide contaminated soils, landfarming, **177:**169
Pesticide contaminated soils, phytoremediation, **177:**169
Pesticide contaminated soils, soil separation techniques, **177:**162
Pesticide contaminated soils, zero-valent iron treatment, **177:**164
Pesticide degradation, veterinary scenarios (data), **180:**37
Pesticide disposal, **177:**123 ff.
Pesticide disposal, bulk quantity FAO guidelines, **177:**135
Pesticide disposal, developing countries, FAO guidelines, **177:**135
Pesticide disposal, incineration, **177:**136
Pesticide half-lives, aquatic, **175:**190–233
Pesticide half-lives, soil concentration effect, **177:**127
Pesticide hydrolysis, (see also Abiotic pesticide hydrolysis), **175:**79, 90
Pesticide hydrolysis, environmental adsorption effect, **175:**166
Pesticide hydrolysis, environmental buffer catalysis, **175:**164
Pesticide hydrolysis, environmental catalysis metal ions/oxides, **175:**178
Pesticide hydrolysis, environmental clay effect, **175:**173
Pesticide hydrolysis, environmental dissolved organic matter effect, **175:**169
Pesticide hydrolysis, environmental micellar effects, **175:**185
Pesticide hydrolysis, environmental solvent effect, **175:**165

Pesticide hydrolysis, kinetics, **175:**81
Pesticide hydrolysis, kinetics mechanisms, **175:**90
Pesticide hydrolysis, metal ion catalysis mechanisms, **175:**178
Pesticide incineration efficiency, **177:**136
Pesticide incineration, incinerator types, **177:**139
Pesticide incineration, muffle furnace temperatures, **177:**137
Pesticide incineration, products of incomplete combustion, **177:**138
Pesticide incineration, temperatures of complete combustion, **177:**137
Pesticide mixtures, disposal problems, **177:**126
Pesticide persistence at high concentrations, **177:**127
Pesticide plastic and paper packaging, incineration, **177:**143
Pesticide potentiation, **172:**39
Pesticide root zone model (PRZM), triclopyr, **174:**34
Pesticide terrestrial toxicity, veterinary use, **180:**63 ff.
Pesticide toxicity groups, honeybee, EPA-defined, **179:**66
Pesticide trace amounts, analytical methods, **175:**80
Pesticide unused sprays, treatment & disposal, **177:**144
Pesticide use (home), house dust contaminants, **175:**17
Pesticide waste constituents, characterization, **177:**132
Pesticide waste, disposal & degradation, **177:**123 ff.
Pesticide waste disposal, Australian regulations, **177:**129
Pesticide waste disposal, dark or photoassisted Fenton's reagent, **177:**158
Pesticide waste disposal, dehalogenation, **177:**151
Pesticide waste disposal, hydrolysis, **177:**150
Pesticide waste disposal, IUPAC recommends, **177:**179
Pesticide waste disposal, oxidation, **177:**152

Pesticide waste disposal, ozonation, **177:** 152, 154

Pesticide waste disposal, photocatalytic oxidation with titanium dioxide, **177:** 160

Pesticide waste disposal, photolysis, **177:** 150

Pesticide waste disposal, recycling, **177:** 129

Pesticide waste disposal, regulatory constraints, **177:**128

Pesticide waste disposal, selecting options, **177:**133

Pesticide waste disposal, strategy, **177:** 131

Pesticide waste disposal, U.S. regulations, **177:**128

Pesticide waste management methods, Australia, **177:**132

Pesticide waste, unique characteristics, **177:**125

Pesticide wastewater, carbon & lignocellulosic sorption, **177:**146

Pesticide wastewater disposal, **177:**143

Pesticide wastewater, evaporative bed disposal, **177:**145

Pesticide wastewater, options for disposal, **177:**145

Pesticide wastewater, ozonation/bioreactor for cleaning, **177:**148

Pesticide wastewater, sorption/composter for cleaning, **177:**148

Pesticide water solubility, dissolved organic matter, **175:**87

Pesticides, abiotic hydrolysis aquatic environment, **175:**79 ff.

Pesticides, affecting fish vertebrae, **172:** 11

Pesticides, aquatic half lives, **175:** 190–233

Pesticides, chiral enrichment processes, **173:**105

Pesticides, chiral list, **173:**90

Pesticides, chlorinated, destruction by K-polyethylene glycol ether (KPEG), **177:**140, 142

Pesticides, disposal of unused stocks, **177:**134

Pesticides, effects on teleosts' calcium/phosphate regulation, **172:**8

Pesticides, half-lives aquatic environment, **175:**190–233

Pesticides, house dust concentrations, **175:**19

Pesticides in fish, Gulf of Gdańsk, **179:**27

Pesticides, levels Gulf of Gdańsk, **179:**19, 24, 26

Pesticides, not recommended for incineration, **177:**139

Pesticides, pure enantiomer advantages of using, **173:**86

Pesticides, racemic disadvantages of using, **173:**86

Pet medicines, environmental pathways, **180:**16

Petroleum crude oil, vertebrate biomonitor effects, **176:**88

Petroleum crude oil, vertebrate rank order & utility index, **176:**94

Petroleum crude oil, vertebrate rank order & vulnerability index, **176:**95

Petroleum crude oil, vertebrate utility index, **176:**89

Petroleum crude oil, vulnerability index, **176:**90

PFAS (perfluoroalkylated substances), **179:**99

PFAS, acute toxicity aquatic organisms, **179:**115

PFAS, biotransformation, **179:**111

PFAS, degradation products, **179:**105

PFAS, ecological risk assessment, **179:** 112

PFAS, ecotoxicity, **179:**111

PFAS emissions, Netherlands, **179:**104

PFAS, environmental fate, **179:**105

PFAS, environmental occurrence, **179:** 108

PFAS, groundwater occurrence, **179:**108

PFAS, LOEC$_{50}$ values aquatic organisms, **179:**112

PFAS, mutagenicity testing, **179:**114

PFAS, occupationally exposed worker levels, **179:**115

PFAS, physicochemical properties, **179:** 106

PFAS soil-repellency illustrated, **179:**103

PFAS, toxicity mechanism, **179:**111

PFAS, uses & applications, **179:**102

PFOA (perfluorooctanoic acid), **179:**100

PFOA, half-life in humans, **179**:114
PFOA, human toxicity, **179**:114
PFOA, peroxisome proliferator, **179**:111
PFOS (perfluorooctyl sulfonate), **179**:100
PFOS, bioaccumulation factor fish, **179**: 110
PFOS, bioconcentration factors, **179**:115
PFOS, ecological risk assessment, **179**:112
PFOS, half-life in humans, **179**:114
PFOS, human toxicity, **179**:114
PFOS, LOEC$_{50}$ values aquatic organisms, **179**:112
PFOS, marine biota contaminant, **179**: 105, 109
PFOS, marine mammals, birds, fish contaminant, **179**:109
PFOS, weakly carcinogenic, **179**:116
pH, chromium redox reaction effects, **178**:97, 99
pH, effect in algal toxicity testing, **178**:28
pH, effect on chromium oxidation, **178**:71
pH effects, phosphate compounds rhizosphere, **177**:29
pH, influences water photochemical transformations, **172**:167
pH rate profiles, abiotic pesticide hydrolysis, **175**:82
Phenanthrene (PAH), from coke production, **174**:13
Phenanthridine, biotransformation products (illus.), **173**:46
Phenolic compounds, house dust contaminants, **175**:14, 26
Phenoxypropanoic acids, chiral herbicides, **173**:90
Phenthoate photoproducts, **172**:204
Phenyl pyrazole insecticides, defined, **176**:2
Phorate, hydrolysis pathways, **175**:137
Phorate photoproducts, **172**:198
Phosalone photoproducts, **172**:208
Phosalone, sublethal AChE inhibition aquatic organisms, **172**:45
Phosphamidon, hydrolytic profile, **175**: 212
Phosphate amendment, lead-contaminated soil remediation, **177**:4
Phosphate, ammonium, fertilizers, **177**:10
Phosphate compounds, acidity equivalent in soils, **177**:28

Phosphate compounds, antagonism effect on mycorrhizal plant infection, **177**: 34
Phosphate compounds, effects on heavy metals in soils, **177**:12, 13
Phosphate compounds, heavy metal immobilization in soils, **177**:18
Phosphate compounds, heavy metal mobilization in soils, **177**:18
Phosphate compounds, heavy metals source in soils, **177**:12
Phosphate compounds, liming value in soils, **177**:28
Phosphate effects on heavy metals, rhizosphere, **177**:29
Phosphate fertilizers, equilibrium dissolution reactions, **177**:11
Phosphate fertilizers, fast-release, **177**:10
Phosphate fertilizers, heavy metal concentrations & sources, **177**:18
Phosphate fertilizers, heavy metal major source, **177**:5
Phosphate fertilizers, slow-release, **177**:10
Phosphate insecticides, hydrolysis mechanisms, **175**:132
Phosphate, monoammonium fertilizers, **177**:10
Phosphate, monocalcium fertilizers, **177**: 10
Phosphate rocks, soil fertilizers, **177**:10
Phosphate-induced heavy metal adsorption, **177**:22
Phosphates, organophosphate photochemistry, **172**:151
Phosphates, reactions in soils, **177**:10
Phosphates, water-insoluble compounds, **177**:10
Phosphates, water-insoluble, effects mycorrihizal plants, **177**:32
Phosphates, water-soluble forms, **177**:10
Phosphonate insecticides, hydrolysis mechanisms, **175**:142
Phosphonates, orgnophosphate photochemistry, **172**:151
Phosphonodithioate insecticides, hydrolysis mechanisms, **175**:142
Phosphonothioate insecticides, hydrolysis mechanisms, **175**:142
Phosphoramidate insecticides, hydrolysis mechanisms, **175**:140

Phosphoramidothioate insecticides, hydrolysis mechanisms, **175**:140
Phosphorescence, defined, **172**:134
Phosphorodithioate insecticides, hydrolysis mechanisms, **175**:136
Phosphorothioate insecticides, hydrolysis mechanisms, **175**:129
Phosphorothiolate insecticides, hydrolysis mechanisms, **175**:134
Phosphorus fertilizers, sources of heavy metals in soils, **177**:4
Phosphorus, role in heavy metal avail to soil-plant system, **177**:1 ff.
Phosphorus-metal interactions in plants, **177**:20
Photoassisted Fenton's reagent, pesticide waste disposal, **177**:158
Photochemical absorption coefficients, organophosphates (table), **172**:149
Photochemical lamps, emission comparisons, **172**:143
Photochemical reactions, actinometry, **172**:142
Photochemical reactions, in atmospheric, **172**:139
Photochemical reactions, in water, **172**:139, 166
Photochemical reactions, light sources, **172**:141
Photochemical reactions, on plant surfaces, **172**:140
Photochemical sensitization, pesticide studies, **172**:190
Photochemistry, defined, **172**:132
Photochemistry, effective radiation spectrum, **172**:137
Photochemistry, irradiation methods, **172**:144
Photochemistry of organophosphorus insecticides, **172**:129 ff.
Photochemistry, OPs (organophosphates), **172**:129 ff.
Photochemistry, pesticides general, **172**:136
Photochemistry, vapor phase described, **172**:145, 147, 168
Photodecomposition, triclopyr, **174**:22
Photodegradation, fipronil, **176**:12
Photodegradation pathways, fenvalerate, **176**:147, 149

Photoenhanced toxicity, aquatic contaminants, **173**:13
Photoenhanced toxicity, azaarenes, **173**:59
Photohydrolysis, methyl bromide, **177**:60
Photolysis, bifenthrin, **174**:71, 72
Photolysis, cyfluthrin, **174**:77, 80
Photolysis, cypermethrin, **174**:87, 90, 91
Photolysis, deltamethrin, **174**:106, 109, 110
Photolysis, esfenvalerate, **174**:121, 122
Photolysis, fenpropathrin, **174**:134, 136
Photolysis, fenvalerate, **176**:146
Photolysis half-lives, pyrethroids, **174**:52
Photolysis, lambda-cyhalothrin, **174**:142, 147, 146, 148
Photolysis, permethrin, **174**:154, 160, 161
Photolysis, pesticide waste disposal, **177**:150
Photolysis, pyrethroids, **174**:61, 62
Photolysis, tralomethrin, **174**:110, 114
Photolysis, triclopyr, **174**:22
Photoproducts, azinphos-ethyl, **172**:212
Photoproducts, azinphos-methyl, **172**:210
Photoproducts, chlorfenvinphos, **172**:153
Photoproducts, chlormephos, **172**:196
Photoproducts, chlorpyrifos, **172**:185
Photoproducts, coumaphos, **172**:192
Photoproducts, cyanophos, **172**:180
Photoproducts, diazinon, **172**:189
Photoproducts, dioxabenzophos, **172**:195
Photoproducts, fenitrothion, **172**:170
Photoproducts, fenthion, **172**:181
Photoproducts fosmethilan, **172**:205
Photoproducts, iodofenphos, **172**:176
Photoproducts, isofenphos, **172**:24
Photoproducts, malathion, **172**:200
Photoproducts, mephospholan, **172**:213
Photoproducts, methidathion, **172**:206
Photoproducts, parathion, **172**:157, 159, 164
Photoproducts, phenthoate, **172**:204
Photoproducts, phorate, **172**:198
Photoproducts, phosalone, **172**:208
Photoproducts, phthalophos, **172**:209
Photoproducts, profenophos, **172**:178
Photoproducts, propaphos, **172**:155
Photoproducts, prothiophos, **172**:202
Photoproducts, quinalphos, **172**:191
Photoproducts, sulprofos, **172**:201

Photoproducts, temephos, **172:**184
Photoproducts, tetrachlorovinphos, **172:**154
Photosynthesis, azaarene photo-products effects on diatom, **173:**61
Phototoxic effects, azaarenes, **173:**53
Phototoxic effects on aquatic organisms, PAHs, **173:**57
Phthalate esters, designated toxic pollutants, **172:**92
Phthalate plasticizers, uses in plastics industry, **172:**88
Phthalates, house dust contaminants, **175:**14, 26
Phthalates, in plastic consumer products, **172:**88
Phthalophos photoproducts, **172:**209
Physical/chemical properties, methyl bromide, **177:**49
Physical properties, pyrethroids, **174:**49 ff.
Physicochemical properties, avermectin, **171:**113
Physicochemical properties, diisononyl phthalate, **172:**89
Physicochemical properties, fenvalerate & esfenvalerate, **176:**140
Physicochemical properties, fipronil, **176:**3, 6
Physicochemical properties, PFAS, **179:**106
Physicochemical properties, pyrethroids, **174:**49 ff.
Physicochemical properties, triclopyr, **174:**20
Physiochemical properties, spinosad, **179:**38
Phytoremediation, gold mine arsenic contamination, **180:**141
Phytoremediation, pesticide candidates, **177:**171
Phytoremediation, pesticide contaminated soils, **177:**169
Phytotoxicity, methyl bromide, **177:**66
Picloram, soil leaching index, **174:**33
Picloram, soil leaching qualities, **174:**30
Pimephales promelas, pentachlorophenol tissue residues, **173:**16
Piperonyl butoxide, antagonism with fipronil, **176:**5

Piperonyl butoxide, house dust concentrations, **175:**19
Pirimiphos methyl, hydrolytic profile, **175:**212
Pirimiphos methyl, sublethal AChE inhibition aquat organ, **172:**45
Planck's constant, abiotic pesticide hydrolysis, **175:**83
Plant nutrition, essential elements defined, **177:**2
Plants, arsenic lethal/sublethal effects, **180:**143
Plants, metal bioavailability in soils, **178:**9
Plants, metal toxicological bioavailability in soils, **178:**10
Plants, PAH toxicity, **179:**76
Plasticizers, world use, **172:**89
Plug-forming bacteria, iron accumulating, **173:**137
Poisoning, arsenic humans, **180:**154
Pollinators, spinosad toxicity, **179:**37 ff.
Pollutant contamination, indoor house dust, **175:**1 ff.
Pollutant risk assessment, vertebrate biomonitors, **176:**67 ff.
Pollutant sources, house dust, **175:**15
Polychlorinated biphenyls (PCBs), house dust contam, **175:**14, 24
Polychlorinated boranes, chiral insecticide, **173:**91
Polychlorinated furans (PCDFs), house dust contam, **175:**27
Polycyclic aromatic hydrocarbons (PAHs), **173:**40
Polycyclic aromatic hydrocarbons (PAHs), **174:**12
Polycyclic aromatic hydrocarbons (PAHs), ecotoxicity data, **179:**73 ff.
POM (particulate organic matter), **175:**2
Pomeranian Bay, most polluted area of Baltic, **179:**4
Portable personal air sampler, house dust, **175:**9
Potassium polyethylene glycol ether (KPEG), destruction of Cl-pesticides, **177:**140, 142
Potentiation, OPs by EBI (ergosterol biosynthesis inhibiting) fungicides, birds, **172:**39

Primisulfuron, hydrolytic profile, **175**:224
Priority pollutants, chromium, **178**:94
Procloraz, ergosterol biosynthesis inhibiting fungicide, **172**:39
Procymidone, hydrolytic profile, **175**:204
Products of incomplete combustion, pesticides, **177**:138
Profenofos, hydrolytic profile, **175**:212
Profenofos, sublethal AChE inhibition aquatic organisms, **172**:44
Profenophos photoproducts, **172**:178
Prometryn, hydrolytic profile, **175**:230
Propachlor, hydrolytic profile, **175**:194
Propanil, hydrolytic profile, **175**:196
Propaphos photoproducts, **172**:155
Propazine, hydrolytic profile, **175**:230
Propetamphos, hydrolytic profile, **175**:220
Propham, hydrolytic profile, **175**:200
Proposed criteria, arsenic environmental protection, **180**:156
Propoxur, house dust concentrations, **175**:19
Propoxur, hydrolytic profile, **175**:200
Propyzamide, alkaline hydrolysis pathway, **175**:106
Propyzamide, hydrolytic profile, **175**:196
Prosulfuron, hydrolytic profile, **175**:224
Prothiophos photoproducts, **172**:202
Protozoa, potential waterborne transmission, **180**:93
PRZM (pesticide root zone model), triclopyr, **174**:34
Pseudokirchneriella subcapitata (alga), algae metal monitoring, **178**:23, 30
Pure enantiomer pesticides, advantages of using, **173**:86
Pydrin®, fenvalerate, **174**:117
Pyrazolate hydrolysis, **175**:160
Pyrazolynate, hydrolytic profile, **175**:232
Pyrethroid chemistry & fate, **176**:137
Pyrethroid hydrolysis, benzoin ester formation, **175**:97
Pyrethroid insecticides, house dust concentrations, **175**:18, 20
Pyrethroid insecticides, hydrolysis, **175**:92
Pyrethroid insecticides, structures & hydrolytic profiles, **175**:192
Pyrethroids, abiotic chemical properties, **174**:60

Pyrethroids, abiotic kinetic analyses, **174**:58
Pyrethroids, adsorption/desorption soil coefficients, **174**:56
Pyrethroids, aerobic soil degradation, **174**:63
Pyrethroids, anerobic soil degradation, **174**:64
Pyrethroids, aquatic half-lives, **174**:52
Pyrethroids, BCFs, **174**:52, 54
Pyrethroids, bioconcentration factors (BCFs), **174**:52, 54
Pyrethroids, biotic chemical properties, **174**:63
Pyrethroids, biotic kinetic analyses, **174**:58
Pyrethroids, chemical names, **174**:66, 73, 82, 95, 117, 125, 139, 151
Pyrethroids, Chemical Abstracts (CAS) regis Nos., **174**:66, 73, 82, 95, 117, 125, 139, 151
Pyrethroids, Class 5 hydrolysis, **175**:99
Pyrethroids, Classes 1 & 2 hydrolysis, **175**:95
Pyrethroids, Classes 3 & 4 hydrolysis, **175**:98
Pyrethroids, enantiomers, **174**:50,
Pyrethroids, endocrine disruption, **176**:152
Pyrethroids, Freundlich adsorption data, **174**:56
Pyrethroids, generation classes, **176**:138
Pyrethroids, half-lives, aerobic aquatic systems, **174**:66
Pyrethroids, half-lives, soil, **174**:65
Pyrethroids, Henry's law constants, **174**:52, 54
Pyrethroids, hydrolysis, **174**:60
Pyrethroids, hydrolysis half-lives, **174**:52
Pyrethroids, K_h, **174**:52, 54
Pyrethroids, K_{oc}, **174**:52
Pyrethroids, K_{ow}, **174**:53
Pyrethroids, K_{rs}, **174**:55
Pyrethroids, mixtures of steroisomers, **174**:50
Pyrethroids, molecular weights, **174**:52
Pyrethroids, octanol-water partition coefficients (K_{ow}), **174**:52, 53
Pyrethroids, photolysis half-lives, **174**:52
Pyrethroids, photolysis, water/soil, **174**:61

Pyrethroids, physicochemical properties, **174**:49 ff.
Pyrethroids, soil adsorption partition coefficients, **174**:55
Pyrethroids, soil half-lives, **174**:52
Pyrethroids, soil sorption, **174**:54
Pyrethroids, synthesis, **176**:141
Pyrethroids, vapor pressures, **174**:52, 53
Pyrethroids, water solubility, **174**:52, 54
Pyrolan, hydrolytic profile, **175**:200

QSARs (Quantitative Structure-Activity Relationships), **173**:1 ff.
Quantitative Structure-Activity Relationships, (QSARs), **173**:1 ff.
Quinalophos, hydrolytic profile, **175**:212
Quinalphos photoproducts, **172**:191
Quinoline, microbial degradation pathway, **173**:43
Quinoline mutagenicity, **173**:63
Quinoline toxicity, **173**:42
Quinoline toxicity, species groups, **173**:49
Quinolones, mechanisms of action, **171**:39
Quizalofop-ethyl, hydrolytic profile, **175**:190

Race, effect on hair trace element contamination, **175**:60
Racemic pesticides, disadvantages of using, **173**:86
Radiation energies, producing homolytic bond fission (table), **172**:138
Ranking methodology, vertebrate biomonitors, **176**:71
Ranking vertebrate biomonitors, environmental contaminants, **176**:67 ff.
Rayonet Photochemical Reactor, described, **172**:145
Reaction quantum yield, defined, **172**:134
Reactive chemicals, tissue residues aquatic organisms, **173**:18
Reactives/inhibitors, CBRs, **173**:5
Reactives/inhibitors, mode of action, **173**:8
Recycling, pesticide containers, **177**:142
Reductases, Cr(VI) to Cr(III), microbial, **178**:118
Regent®, fipronil proprietary name, **176**:6
Remediation, chromate-contaminated soil/water, **178**:77, 141

Remediation, contaminated sediments, **174**:1 ff
Remediation, Cr(VI)-contaminated water, **178**:81
Remediation technologies, chromate-contaminated soils, **178**:79
Remediation technologies, chromium in wastewater, **178**:141
Reproductive effects, DINP, **172**:103
Reproductive effects, fipronil, **176**:36
Reptiles, ChE recovery times OP exposure, **172**:26
Residue analysis, fipronil, **176**:21
Resistance (bacterial) to antibiotics vs disinfectants, **171**:38
Resistant bacteria, via UV/chlorination, **171**:1 ff.
Resonance fluorescence, defined, **172**:134
Resonance structures, hydroxylated triazine & protonated species, **175**:176
Respiratory disease, arsenic-related, **180**:136
RfDs (oral reference doses), chemical warfare agents, **172**:65 ff.
RfDs, house dust contaminant benchmarks, **175**:32
RfDs, interim recommended six warfare agents, **172**:69
Rhizosphere, acidification by phosphate compounds, **177**:29
Rhizosphere, mycorrhizal effects metal tolerance in plants, **177**:31
Rhizosphere, phosphate-heavy metal effects, **177**:29
Rimsulfuron, hydrolytic profile, **175**:224
Rimsulfuron, sulfonylurea bridge contraction, **175**:152
Risk assessment, metal-contaminated soils, **178**:1 ff
Risk assessment models, veterinary medicines, **180**:6
Risk assessment of metals, using algae toxicity tests, **178**:23 ff.
RMS Titanic, see *Titanic,* **173**:117
Rusticle, environmental cost of covert growth, **173**:133
Rusticle, radiographic image (illus.), **173**:130

Rusticle, typical components of dissected rusticle (illus.), **173**:125
Rusticle, word derivation, **173**:123
Rusticles, chemical composition, **173**:124
Rusticles, consortial microbial structures, **173**:121
Rusticles, focused accumulation sites for iron, **173**:129
Rusticles, found on *Titanic* ocean liner, **173**:117 ff.
Rusticles, growth rates, **173**:123
Rusticles, in water wells, **173**:130
Rusticles, injection/extraction wells remediation sites, **173**:131
Rusticles, lab eval corrosive processes (illus.), **173**:124, 131
Rusticles, relevance to maritime industries, **173**:126
Rusticles, underwater pipeline problems, **173**:134

Saccharopolyspora spinosa, spinosad source, **179**:37
Salinity, Antarctic cold waters, **171**:58
Sample digestion, hair trace element analysis, **175**:64
Sample preparation, hair trace element contamination, **175**:63
Sampling methods, house dust, **175**:9
Sampling methods, human hair, **175**:63
Sampling techniques, environmental DINP, **172**:94
Sarin (GB), review summary, **172**:72, 76
Sarin (GB), RfDs & UFs, **172**:68
Scallops, trace metals Antarctica, **171**:83
Scenedesmus quadricauda (alga), Cu toxicity testing, **178**:29, 34, 37
Scenedesmus subspicatus (alga), toxicity testing, **178**:34
Schrader's formula, organophosphate insecticides, **172**:149
Scientific Committee on Antarctic Research (SCAR), **171**:56
Scientific names, terrestrial vertebrates, **176**:70
Scientific organizations, abbreviations, **172**:119
Scoliosis, fish response to xenobiotics, **172**:11

Scout®, tralomethrin, **174**:96
Seals (Weddell), trace metal levels Antarctica, **171**:81
Seaweeds, trace metal monitors Southern Ocean, **171**:73
Sechura phosphate rock, fertilizer, **177**:20
Sediment, bioaccumulation of contaminants, **174**:3
Sediment, contamination remediation, **174**:1 ff.
Sediment contamination, veterinary medicines, **180**:20
Sediment removal, ecological response, **174**:8
Sediment removal, metal concentration changes, **174**:8
Sediment removal, nutrient effects, **174**:8
Sediment removal, PAH declines, **174**:12
Sediment removal, PCB declines, **174**:10
Sediment removal, persistent toxic organics, **174**:10
Selenastrum capricornutum (alga), metal monitoring, **178**:23, 34, 37
Selenium, house dust concentrations, **175**:30
Semivolatile organic compounds (SVOCs), **175**:2
Serum ChE vs brain AChE activity, lizards OP-exposed, **172**:51
Sex, effect on hair trace element contamination, **175**:53
Sheep-dipping chemicals, environmental contaminants, **180**:28
Simazine, hydrolytic profile, **175**:230
Simulated environmental fate, methyl bromide, **177**:70
Skeletal system, xenobiotic effects on teleosts, **172**:1 ff.
Skeletal system, xenobiotic exposure marker, **172**:11
Slime-producing bacteria, chlorine tolerance, **171**:10
Sludge treatment of soils, shifts metals to more mobile forms, **177**:9
Smoking habits, hair trace element contamination, **175**:55
Soil adsorption capacity, dictated by cation-exchange capacity, **177**:6
Soil chromium, chemistry, **178**:59

Soil contamination, veterinary medicines, **180**:22
Soil fumigant, methyl bromide environmental fate, **177**:45 ff.
Soil half-life, spinosad, **179**:39
Soil half-lives, pyrethroids, **174**:52
Soil half-lives, triclopyr, **174**:39
Soil invertebrates, avermectin effects, **171**:120
Soil persistence, triclopyr, **174**:39
Soil quality criteria (SQC), PAHs, **179**:81
Soil quality criteria, algorithm for deriving, **179**:84
Soil sorption coefficients, pyrethroids, **174**:55
Soil surfaces, photochemical studies, **172**:187
Soil-repellence via perfluoroalkylated substance (PFAS), illustrated, **179**:103
Soils, metal extraction methods, **178**:6
Soils, pesticide contaminated cleanup, **177**:161
Soils, source of heavy metals in food crops, **177**:2
Solid-phase speciation, chromium in soils, **178**:62, 64
Solvents, used in photochemical studies, **172**:138
Soman (GD), review summary, **172**:72, 76
Soman (GD), RfDs & UFs, **172**:68
SOS system, in bacteria, **171**:31
SOS system, mutagenesis in bacteria, **171**:32
Spectrofluorimetry, hair trace element analysis, **175**:65
Spectrophotometry, hair trace element analysis, **175**:65
Spectrum, electromagnetic radiation (diag.), **172**:130
Spinosad, acute toxicity insects, **179**:41
Spinosad, alfalfa leafcutter bee toxicity, **179**:42
Spinosad, bumblebee toxicity, **179**:43
Spinosad, chemical structures, **179**:39
Spinosad, commercial trade names, **179**:38
Spinosad, dislodgeable residue toxicity, **179**:44
Spinosad, environmental fate, **179**:39
Spinosad, formulations, **179**:38
Spinosad, hazard evaluation, **179**:40
Spinosad, honeybee toxicity, **179**:42
Spinosad, honeybee toxicity alfalfa residues, **179**:44, 57
Spinosad, honeybee toxicity almond residues, **179**:57, 60
Spinosad, honeybee toxicity avocado residues, **179**:62, 65
Spinosad, honeybee toxicity citrus residues, **179**:59, 61, 63
Spinosad, honeybee toxicity kiwifruit residues, **179**:62, 64
Spinosad, honeybee toxicity strawberry residues, **179**:50
Spinosad, honeybee toxicity tansy phacelia, **179**:54
Spinosad, honeybee toxicity tomato residues, **179**:47
Spinosad, honeybee toxicity treated crops, **179**:58
Spinosad, insect toxicity greenhouse conditions, **179**:48
Spinosad, mode of action insects, **179**:39
Spinosad, nontarget toxicity, **179**:40
Spinosad, physiochemical properties, **179**:38
Spinosad, reduced-risk insecticide, **179**:38
Spinosad, soil half-life, **179**:39
Spinosad, toxicity bumblebee larvae, **179**:51
Spinosad, toxicity honeybee larvae, **179**:53
Spinosad, toxicity of weathered residues, **179**:46
Spinosad, toxicity to pollinators, **179**:37 ff.
Spinosad, vapor pressure, **179**:39
Spinosyns A & D, chemical structures, **179**:39
Spinosyns, metabolites of *Saccharopolyspora spinosa*, **179**:37
Sponges, trace metal monitors Southern Ocean, **171**:74, 75
SQC (Soil quality criteria), PAHs, **179**:81
Standard artificial soils, OECD, **178**:8
Standard test protocols, algae toxicity tests (table), **178**:26
Stanniocalcin, teleost-secreted, **172**:4

Strengite, phosphate fertilizer, **177:**11
Streptomyces avertimilis, source of avermectins, **171:**112
Streptomycin resistant bacteria, **171:**21, 23
Strontium, house dust concentrations, **175:**29
Subchronic toxicity, DINP, **172:**102
Sublethal OP levels, AChE inhibition aquatic organisms, **172:**44
Success®, spinosad, **179:**38
Sulfadiazine, mechanisms of action, **171:**39
Sulfanilamide resistant bacteria, **171:**21
Sulfate-reducing bacteria, rusticle component, **173:**135
Sulfentrazone, hydrolytic profile, **175:**232
Sulfhydryl enzymes, chlorine effects, **171:**6
Sulfide, methyl bromide degradation in soil, **177:**59
Sulfometuron methyl, hydrolytic profile, **175:**224
Sulfonylurea herbicides, chemical structures, **175:**224
Sulfonylurea herbicides, hydrolysis kinetics mechanisms, **175:**144
Sulfonylurea herbicides, hydrolysis pathways, **175:**145
Sulfonylurea herbicides, hydrolytic profiles, **175:**226
Sulfonylureas, Class 1 hydrolysis, **175:**147
Sulfonylureas, Class 5 hydrolysis, **175:**151
Sulfonylureas, Classes 2, 3 & 4 hydrolysis, **175:**150
Sulfur mustard (HD), review summary, **172:**74, 78
Sulfur mustard (HD), RfDs & UFs, **172:**68
Sulfur mustard agents, air exposure limits, **172:**66
Sulprofos photoproducts, **172:**201
Supercritical fluid extraction, house dust contam, **175:**14
Superphosphate fertilizers, soil reactions, **177:**10
Superphosphate single-, fertilizers, **177:**10
Superphosphate triple-, fertilizers, **177:**10
Surface runoff contaminants, veterinary medicines, **180:**20, 29

Surface runoff, triclopyr, **174:**35
SVOCs (semivolatile organic compounds), house dust contam, **175:**2
Synedra sp. (alga), toxicity testing, **178:**34
Synergism, fipronil insecticide, **176:**5
Synthesis, fenvalerate & esfenvalerate, **176:**141

T (sulfur mustard agent), RfDs & UFs, **172:**68
Tabun (GA), review summary, **172:**71, 76
Tabun (GA), RfDs & UFs, **172:**68
Tannery effluents, source of Cr(VI)-reducing microbes, **178:**115
Tannery sludge composition, **178:**75, 76
Tannery sludge, sequential chromium extraction, **178:**65
Tannery waste, major chromium source, **178:**96
Tannery waste sites, chromium problems, **178:**53 ff.
Tanning of leather with chromium, described, **178:**54
TCDD, tissue residues aquatic organisms, **173:**22
TCP, triclopyr metabolite soil adsorption, **174:**29
TCP, triclopyr soil degradate, **174:**25
TDI (tolerable daily intake), DINP, **172:**114
Tebutylazine, hydrolytic profile, **175:**230
Tefluthrin, hydrolytic profile, **175:**192
Teleosts, bone formation, **172:**10
Teleosts, calcium/phosphorus regulation, **172:**4
Teleosts, pesticides effects on calcium/phosphate regulation, **172:**8
Teleosts, xenobiotic effects on skeletal system, **172:**1 ff.
Telomerization, perfluoroalkylethylate production, **179:**101
Temephos, hydrolytic profile, **175:**220
Temephos photoproducts, **172:**184
Temperature, effect in algal toxicity testing, **178:**33
Temperatures of complete pesticide combustion, **177:**137
Teratogenicity, azaarenes, **173:**66
Termiticide, fipronil insecticide, **176:**8

Terrestrial toxicity, veterinary medicines, **180**:63 ff.
Terrestrial toxicity, veterinary pesticides, **180**:63 ff.
Terrestrial vegetation, proposed arsenic criteria, **180**:157
Terrestrial vertebrates, common & scientific names, **176**:70
Test media, algal toxicity testing, **178**:28
Tetrachlorovinphos photoproducts, **172**:154
Tetrachlorvinphos, hydrolytic profile, **175**:212
Tetracycline, mechanisms of action, **171**:39
Tetracycline resistant bacteria, **171**:21, 23
Tetramethrin, hydrolytic profile, **175**:192
Theoretical photochemistry, organophosphates, **172**:130
Theoretical profile shape method, methyl bromide volatilization, **177**:86
Thifensulfuron methyl, hydrolytic profile, **175**:224
Thin-film irradiation, organophosphate photochemistry, **172**:203
Thiofanox, hydrolytic profile, **175**:200
Thiophosphates, organophosphate photochemistry, **172**:156
Tissue residues, contaminants in aquatic organisms, **173**:1 ff.
Titanic, **173**:1985 discovery of shattered hull, **173**:117
Titanic, microbial rusticle growth, **173**:117 ff.
Titanic, rusticle biomass, **173**:128
Titanium dioxide photocatalytic oxidation, pesticide waste disposal, **177**:160
TMP, triclopyr metabolite, soil adsorption, **174**:29
Tolclofos-methyl, hydrolytic profile, **175**:208
Tolerable daily intake (TDI), defined, **172**:114
Tolerable daily intake (TDI), DINP, **172**:114
Total diet, proposed arsenic criteria, **180**:156
Toxaphene, chiral insecticide, **173**:91
Toxic effects, azaarenes range to different species (graph), **173**:51

Toxic waste, steel drum deterioration on ocean floor, **173**:134
Toxicity, aflatoxin-ammonia reaction products, **171**:153, 160
Toxicity, azaarenes, **173**:39 ff.
Toxicity, chromium, **178**:58
Toxicity, chromium to microorganisms, **178**:126, 128
Toxicity comparison, fipronil vs fipronil-desulfinyl, **176**:37
Toxicity, DINP, **172**:97
Toxicity, fipronil mammals, **176**:33
Toxicity mechanism, PFAS, **179**:111
Toxicity, photoenhanced aquatic contaminants, **173**:13
Toxicity tests, algal culture techniques, **178**:27
Toxicological abbreviations, **176**:4
Toxicological bioavailability, metals in soils, **178**:10
Toxicological bioavailability, metals in soils, defined, **178**:1
Toxicological criteria for RfDs, chemical warfare agents, **172**:80
Toys, DINP in manufacturing, **172**:92
Trace elements, house dust contaminants, **175**:28, 30
Trace elements, influencing factors in human hair, **175**:47 ff.
Trace metal concentrations, Antarctic fish, **171**:63
Trace metal concentrations, marine mammals Southern Ocean, **171**:72
Trace metals, Antarctic coastal benthic organisms, **171**:73
Trace metals, Antarctic freshwater ecosystems, **171**:93
Trace metals, Antarctic krill, **171**:59
Trace metals, Antarctica clams/scallops, **171**:83
Trace metals, Antarctica terrestrial ecosystems, **171**:85
Trace metals, coastal benthic invertebrates (Ross Sea), **171**:75
Trace metals in Antarctic organisms, **171**:53 ff.
Trace metals, seals Antarctica, Irish Sea, Finland Gulf, **171**:81
Trace metals, Terra Nova Bay fishes (Ross Sea), **171**:78

Tracer®, spinosad, **179**:38
Trajectory simulation, methyl bromide volatilization, **177**:87
Tralomethrin, anaerobic soil degradation to deltamethrin, **174**:111
Tralomethrin, biotic chemical properties, **174**:111
Tralomethrin/deltamethrin combination experiments, **174**:114, 118
Tralomethrin, hydrolysis, **174**:109
Tralomethrin, hydrolytic profile, **175**:192
Tralomethrin, photolysis in water, **174**:110
Tralomethrin, photolysis on soil, **174**:110, 114
Tralomethrin, physicochemical properties, **174**:52, 95
Tralomethrin, Scout®, **174**:96
Tralomethrin, soil sorption partition coefficients, **174**:105, 107
Transfluthrin, hydrolytic profile, **175**:192
Transport model, methyl bromide, **177**:70, 74
Triasulfuron, hydrolytic profile, **175**:224
Triazine herbicides, hydrolysis kinetics mechanisms, **175**:155
Triazine herbicides, structures & hydrolytic profiles, **175**:230
Triazine ring opening to acetyltriuret, **175**:148
Triazophos, hydrolytic profile, **175**:212
Tribenuron methyl, hydrolytic profile, **175**:224
Tributyltin, house dust concentrations, **175**:27
Tricalcium phosphate, fertilizer, **177**:11
Trichlorfon, hydrolytic profile, **175**:220
Trichlorfon $I_{50}S$, ChE fish, **172**:28
Trichlorfon, phosphonate-phosphate rearrangement, **175**:143
Trichlorfon, sublethal AChE inhibition aquatic organisms, **172**:47
Triclopyr, 2-butoxyethyl ester metabolite hydrolysis, **174**:21
Triclopyr, 2-butoxyethyl ester metabolite photodecomposition, **174**:23
Triclopyr, adsorption/desorption in soils, **174**:28
Triclopyr BEE, see triclopyr 2-butoxyethyl ester, **174**:23
Triclopyr, chemical name, **174**:20
Triclopyr, chemical structure, **174**:20
Triclopyr, compliance with EPED guidelines (table), **174**:43
Triclopyr, degradation pathway (diagram), **174**:25
Triclopyr, environmental fate, **174**:19 ff.
Triclopyr, formulations, **174**:20
Triclopyr, Garlon®, Grazon®, **174**:20
Triclopyr, groundwater contamination, **174**:38
Triclopyr, half-life groundwater, **174**:42
Triclopyr, half-life in plants, **174**:27
Triclopyr, half-life in soils, **174**:39, 42
Triclopyr, half-life in surface waters, **174**:41
Triclopyr, hydrolysis products, **174**:21
Triclopyr, mode of action, **174**:20
Triclopyr, persistence water/sediments, **174**:36
Triclopyr, photodecomposition, **174**:22
Triclopyr, photolysis half-lives, **174**:23
Triclopyr photoproduct, MDPA, **174**:24
Triclopyr photoproduct, oxamic acid, **174**:24
Triclopyr, physicochemical characteristics, **174**:20
Triclopyr, plant metabolism, **174**:26
Triclopyr, soil adsorption, **174**:29
Triclopyr, soil degradation, **174**:24
Triclopyr, soil half-lives, **174**:42
Triclopyr, soil leaching, **174**:28
Triclopyr, soil leaching index, **174**:33
Triclopyr, soil leaching qualities (table), **174**:31
Triclopyr, soil organic carbon desorption effect, **174**:28
Triclopyr, soil persistence, **174**:39
Triclopyr, surface runoff, **174**:35
Triclopyr, TCP as soil degradate, **174**:25
Triclopyr, TMP as soil degradate, **174**:25
Triclopyr, triethylamine salt, **174**:21
Triclopyr, Turflon®, **174**:20
Triclopyr, uses & herbicidal characteristics, **174**:19
Triclopyr, volatilization, **174**:35
Triclopyr-butotyl, hydrolytic profile, **175**:190
Trifluralin hydrolysis, **175**:160
Trifluralin, hydrolytic profile, **175**:232

Triflusulfuron methyl, hydrolytic profile, **175**:224
Triforine, hydrolysis pathway, **175**:161
Triforine, hydrolytic profile, **175**:232
Trout, aflatoxin feeding studies, **171**:163
Turflon®, triclopyr, **174**:20

UFs, used estimating warfare agent RfDs, **172**:70
Uncertainty factor, DINP human hazard, **172**:113
Uncertainty factors (Ufs), defined, **172**:68
Uncertainty factors, used estimating warfare agent RfDs, **172**:70
Universal Waste Rule, pesticide disposal, **177**:129
Unscheduled DNA synthesis, DINP testing, **172**:108
Uranium, house dust concentrations, **175**:30
Urban effect, hair trace element contam, **175**:61
Urea herbicides, hydrolysis, **175**:105
Urea herbicides, structures & hydrolytic profiles, **175**:198
Ureas, alkaline hydrolysis mechanisms, **175**:107
Urine contamination, veterinary medicines, **180**:22
UV disinfection today, **171**:38
UV irradiation, history of use in water treatment, **171**:26
UV irradiation, increasing resistance in bacteria, **171**:1 ff.
UV irradiation, mechanisms of action, **171**:5
UV light dosages, water treatment (table), **171**:27
UV light, mechanisms of damage to bacteria, **171**:29
UV light, used as a disinfectant by nature, **171**:28
UV use and antibiotic resistance in bacteria, **171**:34
UV-tolerant bacteria, **171**:30

Vacuuming, house dust sampling, **175**:9
Vapor phase photochemistry, described, **172**:145, 147, 168
Vapor pressures, pyrethroids, **174**:52, 53

Variscite, phosphate fertilizer, **177**:11
Verbutin, synergism with fipronil, **176**:5
Vertebrate biomonitors, petroleum crude oil effects, **176**:88
Vertebrate biomonitors, ranked for utility, **176**:113
Vertebrate biomonitors, ranked for vulnerability to contaminants, **176**:115
Vertebrate biomonitors, ranking methodology, **176**:71
Vertebrate biomonitors, sensitivity rank, **176**:85
Vertebrate biomonitors, utility rank, **176**:85
Vertebrate biomonitors, vulnerability rank, **176**:85
Vertebrate rank order, persistent organic pollutant effects, **176**:79
Vertebrate toxicity, avermectins, **171**:126
Vertebrate vulnerability to environmental contaminants, **176**:67 ff.
Vertebrates, ease of collection, **176**:71
Vertebrates, geographic occurrence, **176**:71
Vertebrates in biomonitoring environmental contaminants, **176**:67 ff.
Vertebrates, pollutant exposure potential, **176**:71, 73
Vertebrates, pollutant sensitivity, **176**:73
Vertebrates, population resilience, **176**:73
Vertebrates, quantity of existing exposure data, **176**:72
Vertebrates, vulnerability indices, **176**:72
Vertebrates, vulnerability to cholinesterase-inhibiting pesticides, **176**:81
Very volatile organic compounds (VVOCs), **175**:2
Veterinary anesthetics, **180**:10
Veterinary antibiotics, **180**:9, 10
Veterinary antifungals, **180**:10, 12
Veterinary aquaculture medicines, **180**:12, 17
Veterinary drug degradation scenarios (data), **180**:37
Veterinary drugs, aquatic toxicity, **180**:44 ff.
Veterinary drugs, CAS numbers (table), **180**:76
Veterinary drugs, chemical names (table), **180**:76

Veterinary drugs, terrestrial toxicity, **180:** 63 ff.
Veterinary ectoparasiticides, **180:**9, 10
Veterinary endectocides, **180:**9, 11
Veterinary feed additives, **180:**2
Veterinary growth promoters, **180:**2, 12
Veterinary hormones, **180:**10, 12
Veterinary International Cooperation on Harmonisation, **180:**5
Veterinary medicine uses, **180:**9, 15
Veterinary medicines, aquatic toxicity, **180:**43 ff.
Veterinary medicines, CAS numbers (table), **180:**76
Veterinary medicines, chemical names (table), **180:**76
Veterinary medicines, degradation scenarios (data), **180:**37
Veterinary medicines, disposal of unwanted, **180:**16
Veterinary medicines, endocrine-disrupting effects, **180:**73
Veterinary medicines, environment risk characterization, **180:**5
Veterinary medicines, environmental contaminants, **180:**1 ff.
Veterinary medicines, environmental hazards, **180:**36
Veterinary medicines, environmental monitoring data, **180:**18 ff.
Veterinary medicines, environmental occurrence, **180:**17
Veterinary medicines, environmental pathways, **180:**13
Veterinary medicines, environmental risk assessment, **180:**3
Veterinary medicines, exposure assessment models, **180:**8
Veterinary medicines, fate in manure slurry, **180:**32, 34
Veterinary medicines, fate in sediment, **180:**36
Veterinary medicines, fate in soil, **180:**33
Veterinary medicines, fate in surface waters, **180:**34
Veterinary medicines, feces/urine contamination, **180:**22

Veterinary medicines, fish contamination, **180:**24
Veterinary medicines, freshwater contamination, **180:**18
Veterinary medicines, groundwater contamination, **180:**20
Veterinary medicines, marine contamination, **180:**25
Veterinary medicines, metabolism/environmental fate, **180:**31
Veterinary medicines, metabolism of major classes, **180:**32
Veterinary medicines, oestrogenic activity environment, **180:**72
Veterinary medicines, Phase I decision tree, **180:**6
Veterinary medicines, Phase II decision tree, **180:**7
Veterinary medicines, responsible authorities, **180:**2
Veterinary medicines, risk assessment models, **180:**6
Veterinary medicines, sediment contamination, **180:**20.
Veterinary medicines, soil contamination, **180:**22
Veterinary medicines, soil sorption data (table), **180:**35
Veterinary medicines, terrestrial toxicity, **180:**63 ff.
Veterinary medicines, Tier A decision tree, **180:**4
Veterinary medicines, use scenarios, **180:** 2, 9
Veterinary medicines used in UK (table), **180:**10
Veterinary pesticides, aquatic toxicity, **180:**44 ff.
Veterinary pesticides, terrestrial toxicity, **180:**63 ff.
VETPEC model, veterinary medicines, **180:**9
Vinclozolin, hydrolytic profile, **175:**204
Vistula River, nutrient load to Gulf of Gdańsk, **179:**4, 6
Vivianite, phosphate fertilizer, **177:**11
VOCs (volatile organic compounds), **175:**2
Volatile organic compounds (VOCs), **175:**2

Volatilization boundary condition, methyl bromide, **177**:71
Vulnerability index, persistent organic pollutants, **176**:77
Vulnerability index, vertebrate rank order persistent pollutants, **176**:80
Vulnerability indices, vertebrate biomonitors, **176**:73
VVOCs (very volatile organic compounds), **175**:2
VX, review summary, **172**:73, 76
VX, RfDs & UFs, **172**:68

Warfare agents (chemical), oral reference doses, **172**:65 ff.
Waste disposal, pesticide, **177**:123 ff.
Waste treatment of soils, shifts metals to more mobile forms, **177**:9
Wastewater chlorination, MAR-bacterial selection, **171**:22
Wastewater disposal, pesticide contaminated, **177**:143
Water chlorination, increasing resistance in bacteria, **171**:1 ff.
Water chlorine disinfection today, **171**:25
Water, photochemical transformation organophosphates, **172**:166
Water solubility, metal phosphate compounds, **177**:25
Water solubility, pyrethroids, **174**:52, 54
Water UV disinfection today, **171**:38
Water-insoluble phosphates, effects mycorrihizal plants, **177**:32

Waterborne pathogenic protozoa, **180**:94
Waukegan Harbor, PCB-contamination remediation, **174**:10
Waxes, epicuticular effects on OP photodegradation, **172**:182
White-blooded icefishes, Antarctica, **171**:54
Wildlife, exposure organophosphorus (OP) insecticides, **172**:21 ff.
Wildlife risk, fipronil insecticide, **176**:37
Wildlife toxicity, avermectins, **171**:126
Wipe sampling, OSHA, house dust, **175**:10

X-ray fluorescence, hair trace element analysis, **175**:65
Xenobiotic effects on teleosts' skeletal system, **172**:1 ff.
Xenobiotics, house dust concentrations benchmarks, **175**:33

Zero-valent iron treatment, pesticide contaminated soils, **177**:164
Zeta potential, bacterial cells, **171**:6
Zetacypermethrin, see cypermethrin, **174**:82
Zinc, algal toxicity (table), **178**:37
Zinc concentrations, Antarctic fish, **171**:53 ff., 63
Zinc fixation in soils, **178**:5
Zinc, occupational hair levels, **175**:58
Zinc, soil contaminant sources, **177**:3
Zinc, toxicity to earthworms, **178**:11
Zinc, toxicity to springtails, **178**:11

INFORMATION FOR AUTHORS
Reviews of Environmental Contamination and Toxicology
Edited by
George W. Ware

Published by
Springer-Verlag New York · Berlin · Heidelberg · Hong Kong
London · Milan · Paris · Tokyo

The original copy and one good photocopy of the manuscript, and a diskette with the electronic files for the manuscript, complete with figures and tables, are required. Manuscripts will be published in the order in which they are received, reviewed, and accepted. They should be sent to the Editor.

                Dr. George W. Ware
                5794 E. Camino del Celador
                Tucson, Arizona 85750
                Telephone and FAX: (520) 299-3735
                Email: *gware7@aol.com*

1. Manuscript: The manuscript, in English, should be typewritten, double-spaced throughout (including reference section), on one side of 8½ × 11-inch blank white paper, with at least one-inch margins. The first page should start with the title of the manuscript, name(s) of author(s), with the author affiliation(s) as first-page starred footnotes, and "Contents" section. Pages should be numbered consecutively in arabic numerals, including those bearing figures and tables only. In titles, in-text outline headings and subheadings, figure legends, and table headings only the initial word, proper names, and universally capitalized words should be capitalized.

Footnotes should be inserted in text and numbered consecutively in the text using arabic numerals.

Tables should be typed on separate sheets and numbered consecutively within the text in *arabic numerals;* they should bear a descriptive heading, in lower case, which is underscored with one line and starts after the word "Table" and the appropriate arabic numeral; *footnotes in tables* should be designated consecutively within a table by the lower-case alphabet. *Figures* (including photos, graphs, and line drawings) should be numbered consecutively within the text in arabic numerals; each figure should be affixed to a separate page bearing a legend (below the figure) in lower case starting with the term "Fig." and a number.

To facilitate production, authors are strongly encouraged to submit their manuscripts (including figures and tables) in electronic form on diskette. Manuscripts may be submitted in DOS, Windows, or Macintosh format (but not UNIX) using popular word processing software (e.g., WordPerfect, Microsoft Word) or they can be saved as ASCII files. Tables can be prepared likewise or can be submitted as spreadsheets (e.g., Microsoft Excel). Figures may also be submitted electronically using such programs as Adobe Illustrator and Adobe Photoshop. The rules cannot be narrower than .25 pt. Gray screens should not be paler than 15% and should not be darker than 60%. Screens meant to be differentiated from one another must differ by at least 15%. Files can be saved both in .tif or .eps format only. Figures should be saved in both their original application and as PostScript files. Scanned reproductions of black and white art should be provided as 300 ppi TIFF files. Scanned color illustrations should be provided as TIFF files scanned at a minimum of 300 ppi with a 24-bit color depth. Please provide tables and figures in their own files, not to be included in text files. Authors with questions regarding electronic preparation of their manuscripts are encouraged to contact Jenny Wolkowicki at Springer-

Verlag via phone (212–460-1732), FAX (212–533-5977), or Internet (*JENNYW@ SPRINGER-NY.COM*).

2. Summary: A concise but informative summary (double-spaced) must conclude the text; it should summarize the significant content and major conclusions presented. It must not be longer than two 8½ x 11-inch pages. As a summary, it should be more informative than the usual abstract.

3. References: All papers, books, and other works cited in the text must be included in a "References" section (*also double-spaced*) at the end of the manuscript. If comprehensive papers on the same subject have been published, they should be cited when the bibliographic citations extend farther back than to these papers.

All papers cited in the text should be given parentheses and alphabetically when more than one reference is cited at a time, e.g. (Coats and Smith 1993; Holcombe et al. 1995; Stratton 1999), except when the author is mentioned, as for example, "and the study of Roberts and Stoydin (1991)." References to unpublished works should be kept to a minimum and mentioned only in the text itself in parentheses. References to published works are given at the end of the text in alphabetical order under the first author's name and chronologically, citing all authors (surnames followed by initials throughout; do not use "and") according to the following examples:

Periodicals: Name(s), initials, year of publication in parentheses, full article title, journal title as abbreviated in the "ACS Style Guide: A Manual for Authors and Editors" of the American Chemical Society, volume number, colon, first and last page numbers. Example:

Leistra MT (1990) Distribution of 1,3-dichloropropene over the phases in soil. J Agric Food Chem 18:1124–1126.

Books: Name(s), initials, year of publication in parentheses, full title, edition, volume number, name of publisher, place of publication, first and last page numbers. Example:

Gosselin R, Hodge H, Smith R, Gleason M (1986) Clinical Toxicology of Commercial Products, 4th Ed. Wilkins-Williams, Baltimore, MD pp 119–121.

Work in an edited collection: Name(s), initials, year of publication in parentheses, full title. In: name(s) and initial(s) of editor(s), the abbreviation ed(s) in parentheses, name of publisher, place of publication, first and last page numbers. Example:

Metcalf RL (1978) Fumigants. In: White-Stevens J (ed) Pesticides in the environment. Marcel Dekker, New York, pp 120–130.

Abbreviations

A	acre	min	minute(s)
bp	boiling point	M	molar
cal	calorie	mon	month(s)
cm	centimeter(s)	ng	nanogram(s)
d	day(s)	nm	nanometer(s)(millimicron)
ft	foot (feet)	N	normal
gal	gallon(s)	no.	number(s)
g	gram(s)	od	outside diameter
ha	hectare(s)	oz	ounce(s)
hr	hour(s)	ppb	parts per billion (µg/kg)
in.	inch(es)	ppm	parts per million (mg/kg)
id	inside diameter	ppt	parts per trillion (ng/kg)
kg	kilogram(s)	pg	picogram(s)
L	liter(s)	lb	pound(s)
mp	melting point	psi	pounds per square inch
m	meter(s)	rpm	revolutions per minute
m^3	cubic meter(s)	sec	second(s)
µg	microgram(s)	sp gr	specific gravity

μL	microliter(s)	sq	square (as in "square m")
μm	micrometer(s)	vs	versus
mg	milligram(s)	wk	week(s)
mL	milliliter(s)	wt	weight
mm	millimeter(s)	yr	year(s)
m*M*	millimolar		

Numbers: All numbers used with abbreviations and fractions or decimals are arabic numerals. Otherwise, numbers below ten are to be written out. Numerals should be used for a series (e.g., "0.5, 1, 5, 10, and 20 days"), for pH values, and for temperatures. When a sentence begins with a number, write it out.

Symbols: Special symbols (e.g., Greek letters) must be identified in the margin, e.g. $A=\beta b/2\lambda$ [beta]   [lambda]

Percent should be % in text, figures, and tables.

Style and format: The following examples illustrate the style and format to be followed (except for abandonment of periods with abbreviations):

Sklarew DS, Girvin DC (1986) Attenuation of polychlorinated biphenyls in soils. Rev Environ Contam Toxicol 98:1–41.

Yang RHS (1986) The toxicology of methyl ethyl ketone. Residue Rev 97:19–35.

References by the same author(s) are arranged chronologically. If more than one reference by the same author(s) published in the same year is cited, use a, b, c after year of publication in both text and reference list.

4. Illustrations: Illustrations may be included only when indispensable for the comprehension of text. They should not be used in place of concise explanations in text. Schematic line drawings must be drawn carefully. For other illustrations, clearly defined black-and-white glossy photos are required. Should darts (arrows) or letters be required on a photo or other type of illustration, they should be marked neatly with a soft pencil on a duplicate copy or on an overlay, with the end of each dart indicated by a fine pinprick; darts and lettering will be transferred to the illustrations by the publisher.

Photos should not be less than 5x7 inches in size. Alterations of photos in page proof stage are not permitted. *Each photo or other illustration should be marked on the back, distinctly but lightly, with a soft pencil, with first author's name, figure number, manuscript page number, and the side that is the top.*

If illustrations from published books or periodicals are used, the exact source of each should be included in the figure legend; if these "borrowed" illustrations are copyrighted by others, permission of the copyright holder to reproduce the illustrations must be secured by the author. Permissions forms are available from the Editor and upon completion by the original publisher should be returned to the Editor.

5. Chemical Nomenclature: All pesticides and other subject-matter chemicals should be identified according to *Chemical Abstracts*, with the full chemical name in text in parentheses or brackets the first time a common or trade name is used. *If many such names are used, a table of the names, their precise chemical designations, and their* Chemical Abstract Numbers (CAS) *should be included as the last table in the manuscript, with a numbered footnote reference to this fact on the first text page of the manuscript.*

6. Miscellaneous: *Abbreviations:* Common units of measurement and other commonly abbreviated terms and designations should be abbreviated as listed below; if any others are used often in a manuscript, they should be written out the first time used, followed by the normal and acceptable abbreviation in parentheses [e.g., Acceptable Daily Intake (ADI), Angstrom (Å), picogram (pg)]. Except for inch (in.) and number (no., when followed by a numeral), abbreviations are used without periods. Temperatures should be reported as "°C" or "°F" (e.g., mp 41° to 43°C). Because the metric system is the international standard, when pounds (lb) and gallons (gal) are used, the metric equivalent should follow in parentheses.

7. Proofreading scheme: The senior author must return the Master Set of page proofs to Springer-Verlag within one week of receipt. Author corrections should be clearly indicated on the proofs with ink, and in conformity with the standard "Proofreader's Marks" accompanying each set of proofs. In correcting proofs, new or changed words or phrases should be carefully and legibly handprinted (not handwritten) in the margins.

8. Offprints: Senior authors receive 30 complimentary offprints of a published paper. Additional offprints may be ordered from the publisher at the time the principal author receives the proofs. Order forms for additional offprints will be sent to the senior author along with the page proofs.

9. Page charges: There are no page charges, regardless of length of manuscript. However, the cost of alteration (other than corrections of typesetting errors) attributable to authors' changes in the page proof, in excess of 10% of the original composition cost, will be charged to the authors.

If there are further questions, see any volume of *Reviews of Environmental Contamination and Toxicology* or telephone the Editor (520–299-3735). Volume 159 is especially helpful for style and format.